Date Due

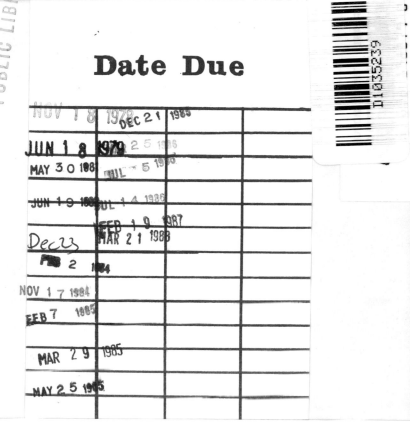

NOV 1 8 1978	DEC 2 1 1985		
JUN 1 8 1979	2 5 1986		
MAY 3 0 1981	JUL 5 1986		
JUN 1 9 1986	JUL 1 4 1986		
	FEB 1 9 1987		
Dec 23	MAR 2 1 1988		
FEB 2 1984			
NOV 1 7 1984			
FEB 7 1985			
MAR 2 9 1985			
MAY 2 5 1985			

3

Freshwater Pollution, Canadian
Style

Canadian Society
of Zoologists

*Environmental Damage
and Control in Canada*
M. J. Dunbar, GENERAL EDITOR

1. **Environment and Good Sense**
   M. J. Dunbar

2. **A Citizen's Guide to Air Pollution**
   David V. Bates

3. **Freshwater Pollution,
   Canadian Style**
   P. A. Larkin

4. **People Pollution**
   Milton M. R. Freeman

P. A. Larkin
FRESHWATER POLLUTION, CANADIAN STYLE

Illustrations by
H. E. Hahn

SPONSORED BY THE CANADIAN SOCIETY OF ZOOLOGISTS
McGILL-QUEEN'S UNIVERSITY PRESS
MONTREAL AND LONDON
1974

©McGill-Queen's University Press 1974
ISBN 0 7735 0197 5 cloth; ISBN 0 7735 0208 4 paper
Library of Congress Catalog Card No. 73-94316
Legal Deposit Second Quarter 1974
Printed in Canada

CONTENTS

PREFACE   ix
ACKNOWLEDGEMENTS   xiii
1. The Nature of Lakes and Streams   1
    Lakes   1
    Rivers and Streams   10
2. Canada's Inheritance   15
    Lakes of the Canadian Shield   17
    The Lakes of the Parkland Zone   21
    Saline Lakes   23
    The Mountain Lakes   26
    The Rivers and Streams   27
3. The Pollution of Lakes and Streams   31
4. Organic Pollution   38
5. Toxicants   45
    Gross Pollutants   45
    Heavy Metals and Mining   47
    Mercury Pollution   49
    Oil Pollution   53
    Herbicides and Pesticides   55
    Pulp-Mill Wastes   59
    Other Industrial Pollutants   63
6. Eutrophication   66
7. Wastes from Power Production—Heat and Nuclear Materials   70
    Waste Heat   70
    Nuclear Waste   73
8. Land Use, Water Use, and Pollution   76
APPENDIX The Canada Water Act   91
REFERENCES   112
INDEX   127

CONTENTS

FIGURES

Between pages 50 and 51

1. Diagrammatic representation of the various factors that influence the characteristics of lakes
2. Production and decomposition zones in lakes
3. Temperature cycle in temperate lakes
4. Influence of lake depth on relative sizes of upper and lower zones
5. Aerial photograph of Hume River, Northwest Territories
6. Lakes of northern Canada
7. *Pontoporeia affinis* and *Mysis relicta*
8. Profile of the Great Lakes
9. Mountain lakes in southeastern British Columbia
10. *Ephydra* and *Artemia*
11. Major drainage regions of Canada
12. Common method of describing toxicity of a compound to an organism
13. Effect of toxicity of zinc sulphate on rainbow trout in waters of different hardnesses
14. Changes in stream conditions downstream from a discharge of sewage effluent
15. Flow diagram of primary, secondary, and tertiary sewage treatment
16. Typical tailings pond
17. Industrial sources of mercury pollution in Canada
18. Diagram indicating the persistence of some common insecticides in the soil
19. Changes in the chemical content of Lakes Erie and Ontario

20. Changes in the fisheries catch from Lakes Erie and Ontario
21. Atomic energy establishments in Canada
22. Effect of fine sediments on the emergence of salmon from their eggs
23. Typical log jam on a salmon-spawning stream
24. Effects of dams on young salmon
25. NAWAPA scheme for water diversion in North America
26. James Bay Hydroelectric Power Development Project

PREFACE

This book describes some of the characteristics of lakes and streams, using examples from Canada, and then outlines a number of things that can happen to lakes and streams when they are influenced by man. This chronicle of Canadian lakes and streams is fairly long—it takes up the first third of the book—intentionally, because it is not generally realized that lakes and streams are widely different in various parts of Canada. There is no single or simple way of approaching pollution problems, and it is only through understanding the background of the biology of lakes and streams that intelligent plans for the future can be developed. The objectives of this book are to help make Canadians better informed about their magnificent freshwater inheritance; to alert them to some present problems and dangers of pollution in their lakes and streams; and to advise them of some of the principles of wise freshwater management.

Lakes and streams are particularly important to Canada's future. Clean, fresh water has always been indispensable to human existence, and is a resource of great value. To date, we have only begun to develop the potential of our great freshwater reserves. As yet, we have not made all the mistakes that others have but we have committed many. Our future growth may well be largely tied to our ability to make fewer mistakes in the years to come, and, by wise management, to aspire to the most constructive and environmentally sound development of a truly great natural resource.

Fresh water plays a key role in the biology of man. None of us can live for long without water to drink. Freshwater fish are a strategically placed staple food in many parts of the world, and a source of recreation wherever they occur. Lakes and streams have been the key to exploration and trade, particularly in Canada. Lakes and streams are chosen sites for man's major cities—they make life possible in most of the habitable parts of the world; they have been the focal points of many civilizations. It may well be these bodies of water provide a special touch of environmental quality in the minds of men—civilization is frequently characterized by special attention to the preservation and creation of quiet ponds and gentle streams. Increasingly, man has been able to control the traits of lakes and streams and to direct the flow of water to the sea. Dams and reservoirs and aqueducts were first built thousands of years ago to enhance the natural patterns of flow. Today, thousands of structures regulate the fresh waters of the world. The storage volume of man-made reservoirs is now sufficient to impound one-tenth of the world's freshwater run-off. Human activities of other kinds detract from natural storage of water. Clearing of forests may much reduce the capacity of the land to soak up and hold water. As a consequence, although the total run-off may be greater after the removal of trees, the natural regulated flow may be replaced by dry stream beds that burst occasionally into heavy floods. As a manipulator of the hydrologic cycle, man is becoming a major factor.

Along with this control over quantities of water flow, man is increasingly influencing the qualities of water. Most of these effects are not planned with enhancement in mind; indeed, many are not planned at all, but are the composite of a great many thoughtless and inadvertent acts. Every discharge of sewage into a lake or stream, every careless disposal of waste materials, has some effect on water quality. The addition of any substance changes the chemistry of fresh water and, in sufficient amounts, will prove harmful to the natural biology. When this happens, we call it pollution—a defilement of the natural state of a lake or stream.

Until recently the scale of human activity was not sufficient to make many of these harmful effects abundantly evident. Western Europe experienced pollution in many of its rivers during the industrial revolution; the same occurred in the latter part of the last century in the more industrialized parts of North America. For Canada, although there were many isolated cases of pollution of local lakes and streams, there was no substantial concern about freshwater pollution until the period of growth after the Second World War. In such a large country, with such a large endowment of fresh water, the deterioration of a lake here or a stream there seemed to be of little consequence. In the past two decades, with the increased tempo of resource development, and the concentration of people into progressively larger urban areas, it has become apparent that water pollution is now a problem where most Canadians live. And, in the rural areas and hinterland, the scale of activities and the new technologies of resource extraction have combined to provide the potential for "instant pollution" of major freshwater environments.

It is impossible to document completely the many instances of water pollution that have occurred to date in Canada. Many have never been recorded, and have long since been discontinued and rectified by natural processes. Even the contemporary scene is difficult to describe comprehensively. New problems of water management and pollution are created almost daily; some, at least, are almost as quickly detected and, where necessary, remedied by the actions of responsible government agencies. Much potential pollution is prevented by abatement measures built into the design of new industrial plants. Some chronic offenders are the subject of phasing-out operations that will gradually lead to higher water quality. But, for all of these circumstances, it is obvious that we are collectively losing ground in maintaining water quality. Where some areas are cleaning up, others are continuing to accumulate problems. The best assessments are those that are made on local and regional levels. If each part of Canada were pollution-free, and if in each region there were wise use of land and water resources, then there would be little cause to worry about national statistics.

In writing this small book, I have been much reminded of the group of fellows in the fisheries division of the British Columbia Game Commission in the early 1950s, and of their counterparts in the other provinces of Canada. They were dedicated to environmental protection then, and gained many of their perspectives from an older generation of game wardens and fisheries officers. It is my hope that their points of view will prevail in the future and that this book will help them in their cause.

March 1973   P. A. LARKIN

ACKNOWLEDGEMENTS

Dr. T. G. Northcote was an invaluable critic. He caught many errors and omissions in the first draft. Robert Scott was very helpful in his perusal and summary of much of the pollution literature. He will recognize the parts that were "cribbed" from his notes. Gordon Ennis gathered the references to Canadian scientific studies on pollution. Max Dunbar, as the instigator of this Canadian Society of Zoologists' series, gave necessary prods and encouragement, and useful advice and criticism. Many students and faculty at the University of British Columbia were helpful in drawing my attention to pieces of literature or in explaining things for me, and I am indebted to them for their interest and encouragement.

ACKNOWLEDGEMENTS

chapter one

THE NATURE
OF LAKES
AND STREAMS

LAKES

To understand the effects of pollution on fresh water, it is first necessary to have a knowledge of the characteristics of lakes and streams, and, particularly, to appreciate the wide variety in the kinds of lakes and kinds of streams. When pollution occurs, it is a consequence of the natural condition of the water receiving the pollutant as well as of the attributes of the pollutant itself. There is no single, useful, standard set of rules that can specify what amounts of what substances can be added to lakes or streams without risk of pollution, and this is so because lakes and streams are so widely different in character.

Lakes are particularly diverse, and they reflect the total impact of their geographic locations. Figure 1 was originally designed by Dr. D. S. Rawson of the University of Saskatchewan, and is a widely used illustration of the multiplicity of forces that make a lake what it is and imply that, however similar two lakes may seem to be, no two are ever quite the same.

The geological formation underlying a lake basin and the surrounding topography determine the quantities and kinds of minerals and sediments that flow into a lake. Just as the soils of some areas are rich, or saline, or acid, so are the waters of various lakes widely different in the basic environment that they provide for plant and animal life. For example, Little Manitou Lake, near Watrous, Saskatchewan, has a salinity of 120,000 parts per million, about four times saltier than the ocean. By contrast, many of the small lakes of the British Columbia coast have less than 25 parts per million of dissolved minerals, and their waters can be safely used to "top up" a car battery. Between these extremes, the great majority of lakes have a dissolved mineral content in the range of 50 to 1000 parts per million. In the north, on the slowly dissolving granitic rocks of the Canadian Shield, lake waters are generally low in dissolved minerals and in the acid range of pH. On the sedimentary deposits of the prairies, lakes are relatively high in dissolved minerals and are alkaline.

The *topography* of a land area influences the characteristics of lakes in two ways: its effects on drainage and on the shape of the lake basin. For example, the mineral content of lake waters is partly dependent on the drainage into a lake. Thus, at the higher altitudes at the top of a drainage system, streams and lakes usually contain only a small quantity of dissolved minerals, reflecting the small size of the drainage area and the short distance a stream has run over the substrate that it slowly dissolves. Further downstream, as a consequence of the dissolving of minerals of the substrate and of the evaporation of water, particularly from lake surfaces, the mineral content of lakes and streams increases. Thus, because their drainages come from a large range of geological formations, major rivers tend to have relatively similar mineral content, whereas small lakes and streams may have unusual characteristics that reflect the homogeneity and small size of their drainage systems.

Topography also determines the shape of a lake basin, which in turn affects almost every facet of its biology. Thus, a large, deep lake such as Lake Superior has many characteristics that are fundamentally different from those of a small, shallow lake on the prairies. A long, thin, deep lake with several arms is strikingly different from one in a saucer-shaped basin.

To understand why the shape of the basin is so important to lake biology, it is necessary to consider the basic processes of production and recycling of nutrients that make a lake an environment for plants and animals. The primary production by plants takes place only to the depths where light penetrates (figure 2): much is performed by rooted aquatic plants around the lake. Thus, the longer the shoreline, the greater the production of vegetation. A lake that has many islands and an irregular margin is more fruitful than one that is almost circular. Additionally, the larger a lake, the smaller is the ratio of shoreline to area, so big lakes, on this count, are relatively less productive than little ones. A lake with broad, shallow beaches has a more fertile shoreline than a lake with steep rocky shores.

Depth of a lake also influences production. The greater amount of plant growth usually occurs in the surface waters of the open lake where phytoplankton (minute drifting plants) carry on the photosynthesis that traps the sun's energy and produces organic matter. A deep lake and a shallow lake of equal area receive the same amount of light from the sun and therefore should have the same production. But plant growth also requires a continuous supply of mineral nutrients, and this can only be provided by recirculation. In a shallow lake, strong winds may stir the water from top to bottom, whereas in a deep lake the force of the wind is insufficient to mix the upper layer with the lower. Thus, after the first spring and summer growth of phytoplankton, which dies and sinks to the bottom, the mineral nutrients may be depleted. In general, then, deep lakes have less production than shallow lakes. There are other, more subtle factors involved in the effects of lake size and shape on productivity, but usually the most fertile lakes are those that are shallow and irregular in shape, whereas the least fruitful are those that are large and deep.

Climate is the final major factor influencing lake biology. If the growing season is long, there is much greater opportunity for a succession of plant forms and for a sequence of "crops." In northern latitudes the shortness of the growing season is compensated in some degree by the greater length of summer days, but, because of the cold temperatures, the more northerly the latitude the less the annual production. In high alpine and high arctic lakes, the ice-free period may be extremely short—perhaps less than six weeks—and, though there may be some photosynthesis under the ice in the spring, the lake waters are open for a major burst of production for only one or two months each year. On the prairies the ice-free period may be six to eight months and, with sustained irradiation, lake surface waters may warm up to temperatures of more than 20°C (68°F) at which all biological processes are much accelerated. In a few parts of Canada lakes may be ice-free for the whole year; but, again, the cold temperatures and limited ligh⁺ intensities virtually confine production to an eight- or nine-montḥ period.

The amount of wind is another element of climate that has major effects on lake structure and production, because the wind does the work that is necessary to stir the waters and to mix the heat from the sun to greater depths. In temperate climates, such as those of much of Canada, lakes have an annual cycle of warming and circulation that depends critically on the wind. In the spring after the ice has gone, the lake has a uniform tempera-ture close to 4°C (39.2°F) from top to bottom (figure 3). This is the temperature at which fresh water is heaviest; when it is either colder or warmer it weighs less and will "float" on the surface above the water that is at 4°C and at maximum density. The surface is soon warmed by the sun and the warmer water, being lighter, floats on the cold water beneath. There is warming at greater depths only if there is a wind to mix the upper warm layer with the deep colder waters. On a clear, cold, early sum-mer day on Great Slave Lake, for example, the upper waters may be warm enough to tempt one to go swimming, but at two feet below the surface the temperature may be only 4°C. It can be a rude shock if you dive in headfirst.

As the summer progresses, the thickness of the upper warm layer increases. By the time it is 30 feet deep, the wind is usually capable of providing enough work to stir just the upper layer. Only a major storm has enough energy to stir some of the warm upper waters into the deep cold body of water below. When this more or less steady state is reached, the lake is said to have an epilimnion, or upper layer; a hypolimnion, or lower layer; and a thermocline, the zone between the two layers in which the temperature changes very rapidly—often as much as a Fahrenheit degree per foot.

At midsummer, the hypolimnion is almost completely uncirculated. It constantly receives a rain of dead plants and animals that disintegrate as they sink slowly to the bottom. The decomposition of this organic matter, both in the hypolimnion and on the bottom, gradually depletes the oxygen dissolved in the water. As a consequence, by late summer, if the volume of water in the hypolimnion is not very large there is virtually no oxygen in the water below the thermocline.

Changes also take place in the epilimnion in midsummer. Although it is warm and receives light, it becomes progressively more unfavourable for photosynthesis because the mineral nutrients in the water have been gradually depleted by the plant growth of the spring and early summer. The nutrients that enabled the early season crops to develop have become locked in the bodies of plants and animals that have died and fallen through the hypolimnion to the bottom. Thus, if the climate remained the same, and if we lived in a perpetual summer, there would be much less top-to-bottom circulation, the surface waters would be less productive, and the deeper waters would be greatly depleted of oxygen. This is precisely what is found in many tropical lakes, such as Lake Tanganyika in Africa.

Lakes can have all degrees of stratification into two layers, and there is great variety in the extent to which they become oxygen-depleted in their lower layers during summer. In lakes that are less than 30 feet deep, it is not uncommon to get warming down to the bottom and some degree of circulation throughout almost the whole summer period. Shallow sloughs are in this category and, since they are continually being recirculated during the summer months, they have high sustained production. If the maximum depth is about 100 feet, and the average depth around 30 feet, there usually is quite sharp temperature stratification, fairly complete removal of oxygen in the lower layer, and high production in the surface layers. These lakes are generally small. When the maximum depth is greater than 100 feet, and the mean depth greater than 30 feet, there is usually some residual oxygen in the hypolimnion even in late summer. In very large, deep lakes, such as Lake Superior or Great Slave Lake, the volume of the hypolimnion is so great that there is little noticeable depletion of oxygen throughout the whole summer (figure 4).

With cooling of the surface waters in autumn, the epilimnion slowly cools so that just before winter the lake may again be uniform in temperature from top to bottom. Autumn winds recirculate the whole lake, recharging the surface waters with mineral nutrients and the deeper waters with oxygen.

If a lake is well protected from autumn winds, or if winter comes with a sudden spell of cold, clear weather, this recirculation may not be accomplished and, as a consequence, the lake may enter the winter period with a substantial oxygen deficit—most of its deeper layers not recharged with oxygen. When this happens, or even if there is complete recirculation but the lake is particularly productive, it may transpire that by spring decomposition has removed the oxygen from all but a very thin layer of water immediately under the ice. In these circumstances, there is no place for the fish to go and many may die of suffocation. In some instances there may be an upper layer of oxygenated water of sufficient thickness that the fish can persist by living up at the surface. But, when the ice breaks up and a spring wind blows, there is a period of a few hours when the layer of surface water is mixed in with the oxygen-depleted layers from the greater depth so that there is temporarily no place that a fish can go where oxygen concentrations are high enough to keep him alive. It is these kinds of situations that cause what is known as "winter kill," not uncommon in lakes of shallow to moderate depth in many parts of Canada.*

More commonly, the circulation in the autumn is sufficient to recharge the oxygen content in the deeper layers to such a level that fish can persist through the winter and the first winds of spring. The spring recirculation, when the lake is again at a constant temperature from top to bottom, is usually much more thorough. Consequently, in most of our temperate lakes the growing season begins again with the mineral nutrients recirculated to the surface and the water saturated with oxygen throughout.

---

*In Canada, the U.S.A., and the U.S.S.R. there have been some successful experiments in preventing "winter kill." Air is pumped into the deep lake waters and creates a circulation that brings deep layers to the surface. Additionally, as the air bubbles rise to the surface, they may help oxygenate the water.

As figure 1 indicates, there are many pathways by which the major agents that influence lake biology are interrelated. The characteristics of a lake are a consequence of the combined effects of all of the factors which can be used to describe its physical setting and its shape. It is for this reason that we say that virtually no two lakes are the same. In the hundreds of thousands of lakes in Canada, there is endless variety as the components of geology, topography, and climate interact in different ways. Characteristically, lakes of a particular district are more similar than those from places that are far removed from each other. For this reason it is common to subdivide large geographic areas into "limnological regions" in which the lakes usually have many features in common. But even this procedure is often of only small convenience, because in some sections there will be several kinds of lakes within a small area. Experts in this field (limnologists) all have stories to tell about the two lakes that looked so much alike and yet which proved, on study, to be so different. Every lake has a character of its own.

The uniqueness of individual lakes is made particularly clear when one looks at the organisms that live in them. In general, and except for bacteria which occur in all parts of a lake, and except for the fish that are usually wide ranging, the plants and animals are divisible into two groups: (1) those that live in the open water (called the plankton); and (2) those that live in and on the bottom. The plankton comprise a wide variety of small plants (usually single-celled or in chains or clusters of single cells) called the phytoplankton, and an equally wide variety of small animals, ranging from protozoa to small crustacea as much as an inch long, collectively called the zooplankton. This whole assemblage of organisms operates more or less as a distinct community in which there are an enormous number of plants, a large number of herbivores (animals that feed on the plants), and a lesser (though still considerable) number of carnivores that feed on the herbivores—a pyramid of numbers with fewer at the higher levels of the food chain.

In virtually every lake there are several hundred species of planktonic organisms, which characteristically appear in greatest abundance at certain seasons of the year, and which perform particular roles in the annual community activity. Just as no two lakes have the same assemblage of organisms, neither is the annual pattern of events quite the same in any two years. A slight shift in the time of spring breakup, a particularly severe storm or two, an epidemic of some fish disease—any such event can cause a small but important change to the natural economy, and a somewhat different annual cycle ensues. It is for these reasons that adequate understanding of the biology of a lake is seldom gained in a single year of study.

Much the same is true of the bottom organisms of a lake. In the very deep cold waters the temperature may not vary either within a year or from year to year. In this cold profundal zone, the seasonal changes in numbers of various kinds of bottom organisms (clams, aquatic "earthworms," small shrimps) may follow much the same pattern from year to year. But elsewhere on the lake bottom fluctuations in numbers of organisms are as common as for the plankton. In the nearshore area, for example, some of the most abundant organisms are the larvae of insects that end their life cycle as flying adults. The "fish flies" which swarm around the lake-shore lights in midsummer are typical enough, and there are many hundreds of species of these small midges in Canadian lakes. Dragonflies and damselflies also have aquatic juveniles which, as wingless nymphs, are among the most common predators on the smaller water organisms. Caddisfly larvae, which live in portable "homes" that they make from small sticks and bits of sand, are another almost universal feature of the shallow water insect fauna. For all of these forms, the success of each year's crop of young depends in part on the weather at the time of the hatch. As a consequence, large fluctuations in numbers are the rule rather than the exception.

Many species of snails, freshwater shrimps, water mites, and aquatic worms also live in, on, and under the rooted aquatic plants of the shore zone. This wet and miniature jungle comprises another distinctive community and, like all other natural communities, shows great complexities in the interrelations between its members and their consequent changes in numbers.

Most of the shore organisms occur only in the shallow waters of a lake and are seldom found beyond the limits of the zone of nearshore plants. This is in part a result of grazing by fish on any organisms that stray from the security of the sheltering vegetation; but, it is also a consequence of the wide temperature fluctuations in the depths between the warm shore zone and the cold depths. Whenever strong winds blow on a lake, they pile up the warm surface waters on the leeward side (sometimes to such an extent that cold water from the depths may well up nearshore on the windward side). When the wind stops, the stage is set for a giant teeter-totter to swing into motion. For the animals that live on the bottom at a depth of 20 to 30 feet, these violent fluctuations (called "temperature seiches") may have the most devastating effects. It can be visualized as being something like living in a shower in which the water changes suddenly and frequently from hot to cold. The organisms of the intermediate depth zones are thus usually a sparse combination of the waifs and strays from shallower or greater depths, plus a few hardy forms that can persist in the rigorous climate.

The fish which occur in lakes may usually be roughly assigned to the planktonic community, the nearshore community, or the bottom community. Thus, the lake herring that feed on zooplankton, and the lake trout that feed on lake herring, are a typical pair of species that are associated with the planktonic assemblage. The common whitefish and various species of suckers are typical bottom feeders. Among the shore dwellers is a great variety of minnows and their predators such as bass, pike, or walleye. These associations of particular species of fish with particular zones are characteristic, but they are not rigid. Most species of fish are to some extent opportunists, and their distribution and abundance in a lake frequently reflect which other species are present (or absent). The young of many species are found very close inshore, but as they mature they usually are offshore and in deeper water. The abundance of various fish species is also subject to annual fluctuation, generated partly by the varying success of the annual crops of young and partly by the interrelations among the fish species.

Collectively, the plankton, bottom organisms, and fish communities are the machinery that make a lake a small functioning biological microcosm. There are, of course, many external influences, and lakes are very much a product of their physical setting, which may vary from year to year. Nevertheless, there is sufficient similarity from year to year that, for each lake, it is possible to construct (1) a budget that describes the way in which the energy from the sun is trapped and converted into many forms of life and (2) a flow diagram that depicts the annual cycles of carbon, nitrogen, phosphorus, and the other elements involved in living processes. When viewed in this way, lakes are described as ecosystems—a physical setting in which a naturally occurring, highly interrelated assemblage of organisms conducts the essential business of perpetuating the fabric of nature.

Like everything else in the world, lakes have a limited life and can be called young or old. From the time a lake basin is formed, it gradually fills in with sediment brought from in-flowing streams and with the slow accumulation of organic matter produced in the lake. Thus, even the largest and deepest lakes may eventually become dry land. If the basin is large and deep when it is formed, then the ecosystem which develops in it is usually described as oligotrophic: cold, deep, low in dissolved minerals, relatively unproductive, with a low ratio of phytoplankton to zooplankton, and a characteristic assemblage of cold-water fish such as trout and whitefish. (The upper Great Lakes are extremes of this oligotrophic type.) As lakes slowly fill, they naturally evolve into eutrophic lakes which are typically warm, shallow, high in dissolved minerals, productive, with a high ratio of phytoplankton to zooplankton, and a complement of warmwater fishes such as crappies, bluegills, and many species of minnows. Many of the smaller lakes of southern Ontario and the southern prairies are eutrophic.

There are many variations on this simple theme of evolution of lakes, but two are worth passing mention. In some areas where soils are especially rich in mineral salts, the eutrophic lake may eventually become a saline lake or slough with only a few salt-loving organisms present. By contrast, if the soil is especially meagre and the drainage into the lake is very low in mineral salts, then the evolution of the lake may eventually proceed to a peat bog, with the soft and acid lake waters stained brown with humic material. Lakes of this kind are common in the Canadian north.

RIVERS AND STREAMS

Rivers and streams are very different from lakes in their biological characteristics. In general, they are much less productive. There is commonly rather limited penetration of light, either because of the overhanging vegetation at the stream margin or because the stream is turbid. Additionally, the substrate and the turbulent flow may be inhospitable for the growth of rooted aquatic plants. For the most part, streams have very little plant production and their biological processes depend on the influx of organic matter from the surrounding land.

Conditions of low productivity are especially true for streams near their headwaters. Where the gradient of a stream is high, the bottom substrate usually comprises only coarse gravel. Characteristically, headwaters show strong fluctuations in stream level. The stream bed may be unstable and alternating floods and droughts may result in the development of braided channels with the formation of transient bars which shift with each freshet. Further downstream, where the gradient is less steep and flows are more constant, streams have more stable channels, their slow-moving reaches may have fine-grained bottom sediments, and the water has a higher mineral content. As a consequence of these many factors, there usually is increasing biological production downstream, particularly from rooted aquatic plants and from algae on the surface of stones. In slow-moving stretches, there may even be some phytoplankton production.

When streams have a very low gradient, they develop meanders which may eventually be cut off from the main flow as oxbow lakes. The Athabasca River from Fort MacMurray to Athabasca Lake, the Slave River from Athabasca Lake to Great Slave Lake, and many of the tributaries of the Mackenzie River provide good examples of meandering streams in an area that has low topography (figure 5). The gradient of the stream generally becomes lower the further one proceeds from the source, and eventually there is almost none in estuarial areas. It is for this reason that estuaries develop their typical patterns of channelling and gradual accretion of deposited silt. In these areas, biological production can become quite high, especially in low marshes that are periodically flooded.

In addition to these broad patterns of change from their sources to their estuaries, streams show many differences which relate to the biological substrate, the climate of the area, and the local topography. In streams, as in lakes, higher productivity is generally associated with higher mineral content. In the Canadian Shield, for example, the streams are acid, low in mineral content, frequently stained brown with humic materials, and, in general, have low productivity. By contrast, streams of the prairies are rich in dissolved mineral content; they are exposed to considerable warming and, in general, have high productivity. In British Columbia, very high rainfall in the coastal area gives streams a low nutrient content and a highly unstable character. Some coastal streams may have practically no total dissolved mineral content—they are virtually sluiceways for large quantities of rain water, especially in the winter. In addition to frequent winter flooding, coastal streams generally have a short and precipitous course to the sea; consequently they are quite inhospitable environments and are not biologically productive.

Variability in flow has a great impact on stream productivity. If the flow is relatively constant, the stream margins are stable and support extensive vegetation which is a rich source of the organic matter that is the basis for stream productivity. If the flow is highly variable, the banks are unstable, there is no vegetation, and the stream may thus be deprived of its major sources of organic matter. When a small stream is flowing in a large scoured channel it is much more exposed to the warming effect of the sun. It is not commonly realized that, for this very reason, stream temperatures can reach levels lethal to fish. Highly variable flow may also result in stream-bank erosion which generates large quantities of fine sediment that may be spread over gravel bars and may convert them into a matrix that is extremely hard and provides a poor environment for stream-dwelling bottom organisms or for spawning fish.

One of the most striking differences between lakes and streams is in their oxygen content. Under natural conditions most streams are fully saturated with oxygen throughout the year because of the turbulence and circulation which constantly brings water to the surface to be recharged with oxygen. For this reason, streams have a substantial capacity to accept organic matter for decomposition. If they are overloaded with organic matter which has a high oxygen demand, their waters may be deficient in oxygen only over a short distance. This characteristic of running waters makes them good natural processors of organic wastes such as sewage, but also renders them susceptible to all of the abuses of excessive pollution.

Streams have another characteristic which has been the subject of considerable study in recent years. On the bottom of most streams there is an assemblage of small animals, largely the larvae of aquatic insects, such as mayflies and stoneflies, that make a living from eating the organic matter which is produced in streams or that falls into them. Most of these organisms have special adaptations to enable them to hold their position on the bottom. Periodically, they may become detached accidentally or they may voluntarily rise from the bottom; they then drift downstream. In almost all streams there is a steady drift of this kind which is snatched at as a passing food supply by the stream-dwelling fish. The drift is most intensive at night. Measurement of the quantities of these drifting organisms is one of the best ways of assessing the productivity of a stream environment. This continued downstream migration would suggest that there would eventually be none in the headwaters. Within a year this occurs for some species, but when they emerge as adults they fly upstream and their offspring begin life in the same places as the parents.

As might be expected, there is a relatively large quantity of drifting food at the outlets of lakes; it is in these areas that stream-dwelling animals may be most abundant. They are surprisingly efficient at removing anything that goes by, so that half a mile below the lake drifting organisms are usually no more abundant than is normal for streams.

Lake outlets may have concentrations of stream organisms that are found only where there is such a rich supply of passing organic matter. For instance, it is not uncommon for the stones at the bottom of the outlet of a lake to be carpeted with *Hydra,* a primitive freshwater relative of coral organisms. In some localities there are concentrations of the freshwater mussel, *Anodonta,* which may be as large as 3 inches in diameter. The latter are of particular interest because they may contain pearls. The upper Mississippi drainage was extensively harvested in the late nineteenth century for freshwater mussels because of the relatively high incidence of good quality pearls.

Stream fishes are of two general types: the permanent residents and the seasonal visitors. The permanent residents in

Canada include trout of various kinds, arctic grayling, mountain whitefish, dace, and cottids; in larger rivers there is a much wider variety, comprising many of the species found in the nearshore and bottom areas of lakes. A notable inhabitant of the largest rivers of Canada is the sturgeon, a very long-lived (over 100 years) and slow-growing species that may attain weights of well over 1000 pounds. In estuarial areas of large rivers there are usually many marine species that tolerate brackish water.

The seasonal visitors give streams their typical piscatorial character and provide a natural reminder that, whereas running water is a well-oxygenated habitat that is highly suitable for development of the eggs, it is not a productive environment rich in food. Most of the fishes that are seasonal visitors to streams are there on spawning migrations. Trout, salmon, suckers, and some species of lampreys are good examples in Canada. For some species, such as the rainbow trout, the existence of a stream suitable for spawning is necessary if a lake is to harbour a permanent self-supporting population.

The seasonal occupation of streams by fish is usually geared to the periods of high water flow. Many smaller streams are intermittent, and it is only for a few months that they provide water for fish to swim in. These watercourses are characteristically devoid of permanent residents and are commonly visited only for spawning or for brief feeding forays by fishes from the lakes or larger rivers into which the streams flow.

In summary, most stream environments are relatively unproductive and are further limited where there is great variability in flow. Streams are best viewed as fragile ecosystems primarily based on the organic matter that they receive from other sources. Most of the organisms that live in streams are specially adapted to the peculiar environmental conditions of the water flow which usually include high oxygen concentrations. It is for these reasons that stream environments are easily influenced by man's activities. It has frequently been observed that streams are the collectors, concentrators, and integrators of all the impacts of man on watersheds. They truly reflect whether we know how to manage our environmental affairs.

chapter two

CANADA'S INHERITANCE

Of all the nations on earth, Canada has the greatest natural endowment of fresh water. The lakes of Canada contain more than one-quarter of the fresh water which is stored on the land surfaces of the world. This huge supply is not a result of receiving more than our share of rain and snow, but rather that the country contains so many basins in which water is naturally stored. Literally hundreds of thousands of lakes are scattered throughout the Canadian Shield in the northern half of the country, their basins being a consequence of the sculpturing of the landscape by the advances of ice sheets in the past hundred thousand years. There are a number of very large basins—the Great Lakes, Great Bear Lake, Great Slave Lake, Lake Athabasca, Lake Winnipeg—all of which are among the world's largest in area, and many of which are among the world's deepest (table 1).

Aside from these major basins and the multitude of northern lakes, there is scattered throughout the whole of the rest of

TABLE 1 Some vital statistics for
Canada's largest lakes and the Great lakes

	Area (square miles)	Drainage Basin (square miles)	Maximum Depth (feet)
Athabasca	3,000	105,980	390
Erie	9,930	32,490	210
Great Bear	11,000	54,590	1,200?
Great Slave	10,502	383,980	2,012
Huron	23,010	72,620	750
Michigan	22,400	67,860	923
Ontario	7,520	34,800	802
Superior	31,820	80,000	1,333
Winnipeg	9,380	380,000	120

Canada a wide variety of small- and medium-sized lakes. For example, in British Columbia there are at least twenty-three thousand, most of which have an area of less than 250 acres. Throughout the prairies, small sloughs and saline lakes are plentiful, and in the aspen parkland and coniferous forest of the central and northern parts of Alberta, Saskatchewan, and Manitoba there are literally several tens of thousands of lakes. Much of Manitoba, Ontario, and Quebec, and the whole of Labrador are in the northern glaciated shield area where lakes may cover as much as 15 to 20 per cent of the total land area. Manitoba alone claims 100,000 lakes, as noted on its automobile licence plates! Even in the sedimentary regions of southern Ontario, the Gaspé Peninsula, and the Atlantic provinces there are thousands of lakes. Only Prince Edward Island is relatively lakeless, with less than one square mile of lake area in its 2200 square miles of land surface.

As indicated in chapter 1, the characteristics of lakes depend a great deal on the surrounding geology, the local topography, and the local climate. Each of the hundreds of thousands of lakes in Canada is thus unique in some respect. Nevertheless, within different parts of the country the lakes tend to have similar characteristics, and they can be treated in large classes. The major limnological regions in Canada are described in the following sections of this chapter.

LAKES OF THE CANADIAN SHIELD

By far the greatest number of lakes in Canada lie in the Canadian Shield, an enormous area of exposed Precambrian formations which has only recently been extensively glaciated. Throughout most of this region the drainage is slow because of the low topography. Lakes may cover as much as 25-30 per cent of the land area (figure 6) and where there are no lakes the terrain is largely composed of rocky outcrops and extensive bogs. The underlying geological substrates are relatively insoluble and, as a consequence, the lakes usually have a low mineral content. Throughout the low-lying bogs there is extensive growth of the moss *Sphagnum* which gives the northern woods their characteristic soft, spongy underfooting. More than 500,-000 square miles of Canada are covered by the typical muskeg plant association of sphagnum moss, labrador tea, blueberry, and black spruce. Throughout this whole area there is every stage in the gradual succession which converts lakes into dry land. Each year the sphagnum bog grows farther out from the edge of the lake until eventually it may cover the surface, even though the basin beneath has not yet been filled in. This stage is called a quaking bog, for when one walks on the surface it "quakes."

The lakes of the Canadian Shield are poorly drained and are commonly stained with humus that is derived from the surrounding bogs. Almost without exception, these lakes are relatively low in dissolved nutrients, are acidic, and for these reasons alone are basically unproductive. The short growing season also implies low productivity. By comparison with lakes in other parts of the world, those in the Canadian Shield are best described as aquatic deserts. Annual natural crops of fish may be in the order of only 2 or 3 pounds per acre per year. As a result, although the lakes of the north are renowned for their attractiveness to sport fishermen, they are not capable of supporting large catches over long periods of time. The very large lake trout which may commonly weigh 30 to 40 pounds are usually fish that are more than 20 years of age. The same slow growth rate is true of the whitefish which occur in virtually all the northern lakes. They are wonderful to eat, but take a long time to grow to harvestable size. It is also common to observe that these slow-growing fish may reproduce only once every three or four years.

It is thus not surprising that, for all of the thousands of lakes in the Canadian Shield, there are very few that have sustained commercial fisheries. For the most part, the many small lakes of the north are best visualized as though they were blueberry bushes many miles apart—although the individual berries are lovely, it is scarcely worth the effort to try to make a meal of them because it is too far between blueberries. It will be important in the future for management of the lakes of the Canadian Shield that we recognize their very low productivity. The successful commercial harvest of their limited crops will depend on relatively light utilization of a large number of lakes.

The circumstances of the recent glaciation of Canada have given the northern lakes some characteristic organisms. As the ice advanced across the Arctic Ocean and Hudson Bay, the land masses were depressed and large quantities of sea water, together with many marine organisms, were shoved ahead of the advancing ice front. Subsequently, as the ice retreated, there were large freshwater lakes formed at the margins of the ice caps. These huge pro-glacial lakes, which once were strung along the whole edge of the continental glaciers covering Canada, gradually receded. A large proportion of our northern lakes are the remnants of these once huge lakes, and they still contain some of the organisms which originated in the Arctic Ocean. One of these, perhaps the best known, is *Mysis relicta* (figure 7). It is shrimplike, about half an inch long, and belongs to a large order of marine organisms which are common as food for marine fishes. They play a similar key role as the food of fishes in Canadian northern lakes. Another glacial relict of almost equal importance in northern Canadian lakes is *Pontoporeia affinis* (figure 7), a small shrimp that occurs in great numbers in the bottom mud in the deeper waters. Where *Mysis* is the main staple of ciscoes and lake trout, *Pontoporeia* is the principal food of the bottom-feeding fishes, such as whitefish.

At great depths in the largest and deepest of the northern lakes there is a small fish which is also a relict of the glacial period. Called *Myoxocephalus quadricornus,* it is similar to the cottids which are so commonly found near the shores of the most southerly lakes, in streams all over Canada, and near the shores on both ocean fronts.

Another consequence of the recent glaciation of the north is the presence of permafrost, permanently frozen soil in which large quantities of ice may be embedded. On the arctic coastal plain, "thaw lakes" are being formed from the ice in the soil as the climate slowly warms. In Alaska some of these thaw lakes have been observed to grow in size by about a yard a year around the edge. This is a particularly novel way for lake basins to form, and it is likely that it has taken place over most of Canada as we thawed out after the last glacial period.

The very large lakes of Canada occur along the margin of the Precambrian Shield and lie partly on the Shield and partly on the sedimentary formations to the south. Great Bear Lake and at least the east arm of Great Slave Lake lie in huge depressions that pre-date the glacial period. The bottoms of both lakes are, at their deepest points, several hundred feet below sea level. For instance, Christie Bay on Great Slave Lake is over 2000 feet deep—its southern shore is an almost sheer vertical rock wall. Lakes Athabasca and Winnipeg were probably sculptured principally by the advancing ice floes. They are remnants of much larger lakes that existed immediately after the retreat of the glaciers and since that time their levels have substantially gone down and their residual basins have been made shallower by the deposition of sediments. For example, the inflow of the Peace and Athabasca Rivers has filled in the western end of Lake Athabasca, at this stage of the process creating a superb marsh for waterfowl. The same is true of the southern end of Lake Winnipeg. The Great Lakes have had a very complex history since their basins were first scoured out by the advancing ice sheet. They are all residues of what were once much larger glacial lakes. Their present drainage area is relatively small considering how large they are as bodies of water. The bottoms are all, except for Lake Erie, below sea level (figure 8).

With the exception of Great Bear Lake, all of the large lakes along the edge of the Precambrian Shield are substantially influenced by the large inflow that they receive from the south. Great Slave Lake, for instance, is both warmed and provided with nutrients by the sizeable input of the Slave River. This is strikingly demonstrated by the 150 ppm (parts per million) of dissolved minerals in the main body of Great Slave Lake, in contrast to the paucity of such nutrients in the east arm of the lake (22 ppm). Similarly, the western end of Lake Athabasca is both warmer and richer in nutrients than the east end because of the influence of the Athabasca and Peace Rivers. Lake Winnipeg receives a warm and nutrient-rich inflow out of the Red River from the south, and via the Saskatchewan River from the west. The Great Lakes, too, contain quantities of dissolved nutrients that reflect the drainage they receive from the south. They also show an increase in dissolved mineral elements from Lake Superior (75 ppm) through Lake Michigan (197 ppm) to Lake Ontario (236 ppm).

It is these inflows of warm and nutrient-rich water that make all of these lakes commercial fishing sites (again, with the exception of Great Bear Lake). Large proportions of Lake Superior, Lake Michigan, and Great Slave Lake are much too deep and cold for high levels of production, but their shallower regions support substantial fish populations.

Throughout the whole of the Canadian Shield and in the large lakes along its margin, the predominant species of fish are the lake trout, the lake herring (or cisco or tullibee), the common whitefish, the northern pike, the northern sucker, and the burbot or ling, all of which are adapted to the relatively cold and unproductive waters of the Canadian north. The same or similar species occur in northern Europe and in the U.S.S.R.

Because of their historical connection to the Mississippi drainage, the Great Lakes contain an additional complement of species which do not occur in the more northern lakes. These include the sauger, chain pickerel, and many species of minnows. Additionally, the Great Lakes have received invasions of several species of fish from the sea by way of the St. Lawrence River. Until about the beginning of this century, Lake Ontario supported populations of Atlantic salmon. The introduction of the sea lamprey into the upper Great Lakes as a consequence of the construction of the Welland Canal is a matter of great concern and common knowledge. By their attacks on fish, the lampreys are said to have virtually eliminated the populations of lake trout in the upper Great Lakes. Lamprey control has proved to be effective but has been expensive and has required a long-term intensive and combined effort by Canada and the United States. Very recently, after many attempts, Pacific salmon (coho) have been successfully introduced in the Great

Lakes. As a consequence of this complex history and of a great many other effects of man (pollution, fishing, soil erosion), the Great Lakes are considerably changed from their natural state and are thus now quite different from the other large lakes along the margins of the Precambrian Shield.

THE LAKES OF THE PARKLAND ZONE

In the wide strip south of the Precambrian Shield, but north of the prairies, there is a group of lakes that are transitional in character. They lie in the band of country that is between the prairies or the hardwood forests to the south and the coniferous forests to the north; since this kind of in-between country is what we usually associate with parks, it is called the "parkland zone." The best examples of these kinds of lakes are found in northern Saskatchewan and Alberta. Although they are highly diverse in their character, they are clearly distinguishable by their intermediate range of dissolved minerals—usually between 200 to 500 ppm, or about 2 to 4 times that of the lakes on the Canadian Shield. It is also distinctive of these lakes that they contain high quantities of carbonate and calcium ions. Their average depths are usually less than 60 feet and they quite commonly exhibit severe oxygen stagnation in the bottom waters in midsummer. Characteristically, they are in a small to intermediate size range. As a consequence of these many features, they are fairly highly productive and support extensive sport and commercial fisheries.

Although the prairie provinces provide the best example of large areas in which this type of lake is common, it also occurs in pockets in other parts of the country. Many of the smaller lakes in the southern part of Nova Scotia and New Brunswick are somewhat similar in character, as are those in southern Ontario where the underlying geological substrate is sedimentary. The lakes of the central interior plateau of British Columbia are also of this type, although they are generally not so productive because of the higher elevation and the shorter growing season.

The parkland lakes are the ones that most Canadians associate with dreams about summer holidays because these are the lakes that often combine sandy shores, moderately warm water for swimming, fairly good fishing, and a scenic setting of mixed forest of deciduous and coniferous trees. These, then, are also the bodies of water beside which so many summer cottages are built, and which, in a variety of ways, can be so substantially influenced by human activities.

Most of the lakes of the parkland belt have received a substantial flora and fauna as a consequence of the many changes of drainage which have taken place since the glacial period. They usually contain a full complement of species of organisms including, if they are deep and cold enough, the glacial relicts *Mysis* and *Pontoporeia*. The deepest of the lakes are characterized by the northern associations of fishes: lake trout, whitefish, suckers, and cisco; but, in the shallower lakes, these species usually disappear and their places are taken by largemouth and smallmouth bass, perch, pike, walleye, and a variety of species of minnows.

The parkland lakes of British Columbia provide a particular and interesting exception. Most of those above 3000 feet elevation in British Columbia do not contain any species of fish naturally. One of the consequences of glaciation was to create a large number of barriers in the drainage systems. As a result, bodies of water in many parts of British Columbia are inaccessible to upstream invasion by fishes, and they have long been called "the barren lakes." This has, of course, created a glorious situation for those who are only interested in fishing for trout. In thousands of lakes in British Columbia the rainbow trout is the only species of fish present—and it has been introduced by man some time in the last 100 years.

There are, of course, disadvantages to this status as "barren lakes." Except for Waterton Lake, which is at the meeting point of Alberta, British Columbia, and the State of Montana, there are no lakes in British Columbia which naturally contain the glacial relicts *Mysis* and *Pontoporeia* that are so important as fish-food organisms. The ice sheets over the Rocky Mountains were separate entities from those which covered the eastern and central parts of Canada, and they did not advance across the ocean; consequently, there has been no source of supply of these organisms. In 1949, *Mysis* was introduced into Kootenay Lake and has since become established there with strikingly beneficial effects on the rate of growth of land-locked salmon (kokanee). *Mysis* has also been introduced into deep, cold lakes in Oregon and California—one of the less well-known Canadian exports!

Another major feature of the parkland lakes in British Columbia is that, where they are accessible to the ocean, they are generally major rearing grounds for sockeye salmon. Shuswap Lake is an excellent example. Every four years it produces a hundred million or more sockeye smolts that provide a return of from two to twenty million adult salmon. Other examples are Babine, Quesnel, and Stuart Lakes. It is a saying among West Coast fisheries scientists that if you are in beautiful country you will find sockeye salmon—which is a testimony to the appeal of the typical Canadian parkland lake setting (figure 9).

On both the Atlantic and Pacific coasts there is a special class of parkland lake which reflects Canada's recent glacial history and the gradual rising of the land as the weight of the glaciers was removed. Powell Lake in British Columbia was, a few thousand years ago, an arm of the sea. It is now a lake 1400 feet deep, the bottom waters of which are still salt and the upper waters of which are fresh. Because the salt water is heavier it will undoubtedly take many thousands of years before the basin is eventually a freshwater lake from top to bottom. On Cape Breton Peninsula Bras d'Or is a salt water "lake" which still has access to the sea, but which is already in the incipient stages of becoming a freshwater lake. There are undoubtedly, on both coasts of Canada, many lakes of this type which as yet have not been adequately studied. It is an interesting aside that in such lakes on the west coast there is a counterpart to the glacial relicts of the north: a small crustacean *Neomysis mercedis*; and in Lake Washington, near the city of Seattle, *Pontoporeia affinis* is found. This is a very curious occurrence of *Pontoporeia*, and to date is the only one along the Pacific coast. It perhaps reflects a curious quirk of the glacial period which may some day be fully understood.

SALINE LAKES

Wherever rates of evaporation are very high and comprise the major "outflow" of water from a lake basin, the mineral salts become so concentrated that the lakes are known as saline–not infrequently far more salty than the waters of the ocean. They occur widely across the southern part of Saskatchewan and Alberta, and in parts of the dry interior of British Columbia. They range in dissolved mineral content from about 1000 to 150,000 ppm. They are usually shallow and, because of the climate of the areas in which they occur, are invariably warm. This combination of high temperature, sunshine, shallowness, and high mineral content makes them extremely productive.

Saline lakes have many unusual characteristics. Where the total mineral content is relatively low, they have rich blooms of blue-green algae; in midsummer they are commonly covered with a green scum that makes them look like a large bowl of asparagus soup. In some instances these algae are toxic and it has been reported that cattle that drink the water may be killed by it. Many of these lakes are potentially highly productive of fish, but they fairly consistently "winter kill." Recently, several of these lakes have been planted with rainbow trout that grow to marketable size within a single summer season, providing an extra crop for the farmers whose lands surround a small saline lake. Fish harvested in the summer from these lakes may have a weedy taste that comes from the presence in the flesh of a substance known as geosmin, which is produced by the blue-green algae. Usually by the late autumn this taste disappears naturally.

Many of the shallow lakes and "sloughs" (in prairie parlance) are important breeding areas for waterfowl. In sequences of years when precipitation is high, there is a blossoming in both the numbers and the size of the many prairie saline lakes, and it is followed by an increase in the numbers of ducks. Many conservation efforts have thus been directed to the preservation and enlargement of saline lakes in Alberta, Saskatchewan, and Manitoba.

When the total dissolved mineral content of saline lakes is in the higher range, they characteristically have a shoreline that is heavily encrusted with precipitated mineral salts. In lakes of this kind, the salt content may make it impossible for fish to live. In the most saline lakes there are only a few species of organisms that can withstand the very high concentrations of salts, but these occur in extraordinary abundance. One of the most typical is the brine shrimp, *Artemia salina*. Another is the brine fly, *Ephydra*. Along the shores of these lakes both *Artemia* and *Ephydra* may be so abundant that every cupful of water contains several hundred (figure 10). The lakes may be so salty that swimming in them is very easy, because even those people with the "heaviest bones" float high in the water.

Over the last 40 years, with changes in the climate on the prairies, many of the saline lakes have gone through long-term changes in their total dissolved mineral content, and this has had a substantial effect on their characteristics. For example, Big Quill Lake near Yorkton, Saskatchewan, had a flourishing whitefish population which gradually disappeared in the late 1930s. The lake was becoming progressively more saline and eventually reached a stage that whitefish eggs could not tolerate. With no young ones coming along, the whitefish population gradually diminished. By the early 1950s, the lake had become sufficiently fresh that whitefish were again able to survive.

Saline lakes may have another characteristic: the deeper layers are sometimes warmer than those at the surface. This curious condition occurs because water that is laden with mineral salts may be heavier than water which is cooler but contains less salts. This situation may also be associated with sulphate bacteria which may occur in very dense layers, at depths where light will penetrate and where hydrogen sulphide is available to them from decomposition in the dense deep waters below. In these, and in several other, respects the saline lakes are quite unusual. Some of them exhibit striking changes in their appearance and in the organisms that inhabit them during the course of the summer season as evaporation makes them progressively saltier. Around their margins they generally have a special association of plants that are adapted to the high levels of mineral salts encrusted on the shoreline. When the wind blows, their waters may quickly generate a broth, and many of them are thus called "soap lakes." Some of them form a surface crust that is broken here and there by patches of water, giving the appearance of a clever design of green circles of various size in a background of glittering white and cream. The saline lakes are in many ways natural curiosities of great scientific interest.

THE MOUNTAIN LAKES

The combination of rugged terrain and recent glaciation has created many special kinds of lakes in the mountainous parts of Canada, particularly in British Columbia, Alberta, and the Yukon. For example, most of the major valleys have been recently scoured by glaciers which flowed down them. The valleys in cross section are noticeably U-shaped with steep sides, and the lakes they contain have only a narrow nearshore zone. The bottoms may be of almost uniform depth over their entire length. The end of the basin is commonly a vast plug of loosely packed material that was deposited as the glacier temporarily hesitated in its retreat. When the glacier again retreated rapidly, another lake basin might be formed, and this pattern is clearly shown in many of the narrow valleys of British Columbia. Trout and Kootenay Lakes form just such a "chain," as do Upper and Lower Arrow Lakes. As a consequence of their long, narrow shape and relatively large inflow, these lakes may have a highly complex physical circulation. Because of their great depth (and many are several hundred feet deep along their length) they are generally unproductive.

The valleys tributary to these long major lake basins have commonly been truncated by the passage of the glacier, and they drain by waterfalls. In the basins above these waterfalls, in magnificent settings of towering mountains, lie small "cirque lakes" that, needless to say, are usually devoid of any natural fish populations, and any trout in them have been introduced by man in the past century. Lakes of the high country are commonly a beautiful translucent whitish-green colour which comes from the presence in their waters of finely crushed rock or "glacial flour." The same colouring may occur in large lakes at lower altitudes wherever they are directly fed from glacial sources.

The uniqueness of lakes is abundantly illustrated in mountainous regions. The tumultuous geology makes each drainage system different from its neighbours. Most mountain lakes in Canada are low in total dissolved minerals because of the shortness of their drainages and the insolubility of the surrounding substrate. Additionally, some lakes are deficient in particular mineral elements; some have a superabundance of particular mineral elements. Some lie in well-sealed basins, and their drainage is visibly accounted for. Others have substantial seepage flow either in or out, or both, and it is often quite mysterious to see a lake level go up without visible inflow, and then down again without visible outflow. The complexities of terrain contribute to many local differences in climate and weather. A lake may be extensively shaded by mountains on its southern shores; another may be particularly exposed to strong winds that funnel through a mountain pass. For these and other reasons, mountain lakes have a great diversity of character, and it is particularly difficult to appreciate their biology without careful study. In southern British Columbia, nine major limnological regions can be distinguished; in varying degrees, each of these can be subdivided. A similar complexity is apparent in the lakes on the eastern side of the Continental Divide in Alberta.

THE RIVERS AND STREAMS

Canada has an enormous number of rivers and streams, the drainage patterns being most complex in the mountains of the west and in the low-lying Canadian Shield. Most of the drainage is to the Arctic Ocean or to Hudson Bay. The Great Lakes and the southern portion of Quebec, as well as all of the Maritime Provinces, drain into the Atlantic. All that portion to the west of the Continental Divide drains into the Pacific. A very small segment of southern Saskatchewan and Alberta drains into the Mississippi (figure 11).

The major rivers of Canada are not the largest in the world. The Amazon, Congo, Mississippi, Yangtse, and Orinoco are all larger than Canada's largest, the St. Lawrence and the Mackenzie, and these, in turn, are over twice the size of any other Canadian rivers. Nevertheless, by international standards, several are worthy of mention: the Fraser River, for example, is comparable in size to the Nile, and the Nelson River is substantially larger than the Rhine. Surprisingly little study has been given to Canada's major rivers, except for knowledge of their volume of flow (and this is lacking or inadequate for most of the northern part of Canada). We thus do not know much about resident populations of fish or stream organisms or productivity.

As might be expected from the climate, Canada's rivers show a strong seasonal element in their biology. Many of the smaller streams are reduced in winter to a small trickle beneath a thick ice cover, and this is also a period of relative inactivity for most of the organisms. There have been very few studies of streams in Canada in winter conditions (and few of lakes in winter as well). Most research has been confined to the summer months.

In general, the streams in Canada follow a pattern of productivity that is related to that of the lakes. In the Precambrian Shield, the multitude of streams follow meandering courses in the low and divided topography, and they are commonly acid and brown-stained. Their season is short and their output low. Their best known product is the myriad of blackflies for which they provide a larval home. In the parkland zone, streams are generally warmer, they contain more mineral nutrients, and they may attain moderate levels of productivity. High yield is only achieved in the southernmost drainages, and even here is not comparable to that of the running waters in the mid to southern regions of the United States.

In the mountainous regions there is a spectacular variety of stream types: from young, wild, and virtually barren streams that plunge down steep slopes to slow, meandering streams in the valley bottoms. Throughout British Columbia the streams in steep terrain are characterized by unstable channels strewn with large gravel and boulders, endless series of cataracts, and only occasional stretches where a mild gradient creates conditions for stream-dwelling organisms. On the coast, heavy winter rainfall and the steep topography extend the youthful geology of streams right down to sea level, and substantial production is only found at the lower elevations in the larger drainage systems.

Whereas Canadian streams are thus usually rather unproductive environments, they nevertheless have a crucial role as spawning sites for many kinds of fishes, particularly the valuable Atlantic and Pacific salmon, and most of the species of trout. Pacific salmon are virtually confined to the drainages into the Pacific, although they do occur in the arctic drainages as far east as the Mackenzie. Atlantic salmon occur along the east coast and in parts of the eastern Arctic. Arctic char and inconnu move in and out of the rivers on the arctic coast. All of these species of fish are known as anadromous fish.* They are born in fresh water, they migrate to the sea, and eventually return to fresh water to spawn. The marine migrations of both Pacific and Atlantic salmon have been the subject of intensive investigation in the last 20 years and it is now known that they may travel distances of several thousand miles offshore. They thus exploit the food resources of the oceans to great distances from land, but their conservation depends upon maintaining the high levels of oxygen in the rivers that harbour their eggs. Moreover, each of the rivers supports a more or less distinct race of fish. The natural rate of dispersal from one river to another is very slow, so the loss of the salmon from a river through a short-term catastrophe may prove to be a long-term disaster.

The rivers and streams of Canada are also the major highways for natural dispersal of fishes. The Continental Divide is the major line of separation of fish faunas in Canada. To the east of the Divide, the fishes belong to the large groups of species that inhabit the Mississippi, Atlantic, and arctic drainages. Since the glacial period, many species of fishes have reinvaded Canada from the south and east, but some of them have also spread north and west from parts of Alaska and the Yukon that were not glaciated.

To the west of the Continental Divide, the fish species belong to a different group that has recolonized along the main river valleys from the south and from the ocean to the west. Thus, on the western side there are no pike or walleye, and their place is taken by squawfish. Different species of minnows occur on the two sides of the Divide. There are many other examples in the total fish fauna, with the western side much less rich in species than the east. There have been some interchanges of fish faunas where alterations in drainages have occurred or where colonization has proceeded along the sea coasts (for instance, there are chum salmon in the Mackenzie drainage), and the processes of mutual exchange continue slowly at the present time. This phenomenon would be of only passing interest if it were not for the fact that interchanges of fish faunas may be fraught with disas-

---

*Eels are found in the rivers and streams that drain into the Atlantic. They are known as catadromous fish. The adults spawn in the ocean (near the West Indies) and the young ones migrate to freshwater streams where they remain until they are adult.

trous consequences, especially for the relatively impoverished fauna of the western drainages. Where there are few species, the addition of a newcomer that has a long history of survival in more intense competitive circumstances may considerably upset the existing balance of numbers between species. Thus, rainbow trout seem to be poor competitors with other species. Additionally, an invading fish species may bring new parasites. The spread of pike, for example, would bring the unsightly parasite *Triaenophorus* to the flesh of Pacific salmon which at present are of such great commercial value. Introduction of exotic species, whether by natural or human agencies, is a tricky ecological business and can be particularly dangerous in mountainous regions where there may be a long history of protective isolation.

chapter three

THE POLLUTION
OF LAKES
AND STREAMS

There are many perspectives in the total picture of water-resources management. Lakes and streams are not only biological systems that man can develop and exploit. They are also avenues for commerce and transportation, potential sources of electric power, natural reservoirs of water for domestic and industrial uses and for agricultural irrigation. They are also the natural sinks that convey materials from the land to the oceans, and, accordingly, reflect the whole pattern of land use of the regions they drain. It is beyond the scope of this book to discuss the complete range of considerations that guide decisions concerning water-resource management. What is attempted in the following chapters is a description of the impact of human activities on the biology of lakes and streams, particularly as they have been observed in Canada, and it begins with a discussion of pollution.

A large proportion of Canada is relatively uninhabited and the lakes and streams are in their natural state. But wherever people are concentrated, or wherever there is extensive human activity, there is abundant evidence of the impact of man on lakes and streams. In the worst situations natural streams have become open sewers which receive uncontrolled discharge of a great variety that reflects the total spectrum of human activity. Thus, many of the streams close to our large metropolitan areas are grossly polluted with human sewage, a wide variety of industrial effluents, street washings, and miscellaneous garbage. The harbour of the City of Toronto is a good example of gross pollution wherein the natural inhabitants of Lake Ontario waters are almost completely absent. Much less obvious, but equally severe, consequences may occur from a single industry which discharges a particular toxic waste. For instance, although a pulp mill or a mining operation may be in a remote area, the impact of its effluent on a stream fauna may be equivalent to the whole gamut of discharges that occur near a large city. It is thus not sufficient to focus attention only on the streams that receive a great number and variety of pollutants, nor to focus on those situations where a single large industry is concerned. In the broad picture of pollution of Canadian lakes and streams, there is reason for concern throughout the whole of the country.

Pollutants in general fall into five categories. First are those which, because of their *organic nature,* overload a stream's capacity to assimilate them. This kind of pollutant includes human sewage, animal manure, vegetable wastes such as potato peelings, forestry wastes such as bark and sawdust—in fact, virtually anything of organic origin that requires oxygen for its decomposition. For these kinds of pollutants the major effect on a stream is to deplete the oxygen concentration in the immediate vicinity of the discharge. In warm climates, the impact may only be evident for a distance of less than a mile or two downstream, but in subarctic and arctic streams, where the flow is as fast but bacterial decomposition is slower, the effect may extend much further downstream. In the broader perspective of the whole stretch of a stream course, this kind of pollution has an enriching effect. It increases the organic base on which stream productivity is founded. It is thus important to realize that small quantities of organic matter discharged throughout the length of a stream course have potential for improving it as an environment. It is only when large amounts are released in local areas that they have such disastrous effects. (This leaves aside, of course, the obvious public-health considerations.)

The second category of pollutant is that which is *directly toxic* to stream organisms, and this includes a great variety of compounds that are produced by industrial complexes. Heavy metals are particularly harmful to fish. Any industrial process or mining operation that discharges large quantities of lead, copper, cadmium, iron, and so on is a source of concern because of the potentially toxic effects. Various organic compounds are similarly toxic. Phenol, which is a common waste material from coking operations and is a common constituent of various kinds of glues, is very toxic to fish. In trace amounts it gives fish flesh the flavour of carbolic soap; in only slightly greater concentrations it is lethal. In the past 50 years the growing complexity of industrial processes has produced a wide variety of chemicals that have various effects on aquatic organisms. In many instances it is not well known what these compounds do to aquatic organisms, but there is sufficient evidence to cause substantial concern. One of the best examples is DDT. Even small quantities of DDT which reach streams from land drainage can cause the mortality of the aquatic insects that are such a large part of the stream-bottom fauna. As is well known, the dead insects are then consumed by fish, in which the DDT may accumulate, eventually causing excessive mortality in the fish eggs. It also impairs the ability of fish to learn, and may have other subtle effects on physiological and behavioural performance.

Some pollutants combine high oxygen demand with direct toxic action. Pulp-mill effluents, for instance, are usually rich in organic matter which depletes oxygen and, at the same time, they contain a variety of organic compounds that are directly toxic. Such mixed effects that may be additive, or that may be particularly severe when two or more toxins combine, are quite common for various kinds of industrial effluents.

A third kind of aquatic pollution is the *enrichment* of lakes and streams by the addition of mineral nutrients that usually limit plankton production. The best known examples in this group are phosphates and nitrates which are also, of course, commonly used as fertilizers on land. To some degree, organic pollution has an enriching effect because it adds these mineral elements as well as carbon. But, commercial fertilizers and detergents are particularly concentrated forms of these nutrients and, when added in sufficient quantity, they may disrupt the natural processes of production in lakes and streams. This disturbance usually takes the form of the production of excessive quantities of plant material, a characteristic of the more fruitful types of natural aquatic environments. For this reason, fertilization of this kind is said to cause "eutrophication"—because of enrichment, the lake takes on some characteristics of eutrophic lakes.

A fourth kind of aquatic pollution is that which involves the *discharge of waste heat.* Many industrial processes require large quantities of cooling water. Steel mills are a good example. Operations of this kind usually discharge water that is considerably warmer than that which they take in. In some situations this can have beneficial effects, but, more commonly, where there is a steel mill there are several other industries and the temperature effects accumulate (one extreme case was reported in the United States in which the temperature of a stream reached 160°F!). Thermal and nuclear power plants may generate large quantities of waste heat and may substantially warm the waters which receive their discharge. Again, if this heat is properly distributed in a large lake, it could have beneficial effects on productivity, although it may not be economically feasible to distribute waste heat efficiently.

Nuclear power plants also involve the possibility of a fifth type of pollution—*radioactive wastes.* The effects of radioactive materials on natural environments are not well known, and have aroused concern that they may cause an acceleration in the rates of mutation and in the incidence of cancer. These kinds of impacts, then, are somewhat more insidious than the direct toxic effects of most industrial effluents.

Pollution regulations are difficult to frame because there is a wide range of effects of different kinds of pollutants on different organisms; these consequences vary with temperature, oxygen concentration, salinity, and other environmental factors. The usual techniques of assessing the effects of a toxic material are to measure, in a laboratory, the concentration of a toxin that kills 50 per cent in a specific period of time (figure 12). This is known as a TL50 or a TLM (Tolerance Limit Median) and is stated with a time period in hours; that is, if the TLM 24 is .05 ppm, it means that 50 per cent of the test animals were dead after 24 hours when the level of toxin was .05 ppm. This method is usually only applied for periods up to 96 hours, so that long-term effects of many toxins are often not well known. More seriously, conditions in nature may depart very widely from those in the laboratory, and, without extensive tests both in the laboratory and in the field, it is difficult to set a single figure as being a safe level for a particular toxic material (figure 13). The fact that various toxicants "work with each other" to produce greater effects than each would have alone is a further complication.

It is one of the major difficulties of pollution investigations, therefore, that the effect of an effluent depends largely on the organisms present and the characteristics of the receiving waters. Thus, discharge of a certain quantity of human sewage into a small stream may have disastrous effects on oxygen levels, but the same amount going into a larger stream could be relatively harmless or even enriching. This has led to the aphorism that the solution to pollution is dilution. Similarly, in the right circumstances, the exhaust of waste heat from industrial operations can increase production if the amounts are small in relation to the body of water that is receiving them. The effect of toxic materials is also largely influenced by the character of the receiving water. For example, heavy metals are far more toxic in soft water than in hard water. Thus, it may be "permissible" to discharge effluent from a mining operation into some streams but not into others.

A good example of the alternately beneficial or harmful effects of pollutants is the addition of phosphates. Discharged into lakes that contain high quantities of total dissolved minerals, particularly sodium, potassium, and calcium, phosphates in even small amounts can generate substantial blooms of blue-green algae. Small quantities of phosphates added to soft-water lakes may not have nearly so great consequences because other nutrient elements are in such short supply that the phosphates alone cannot stimulate algal growth.

For all of these reasons, many kinds of pollution are best viewed as an excessive and harmful discharge of materials and substances into fresh water. For a very wide variety of pollution problems, modification of the effluent or redistribution may bring about amelioration. In the majority of instances of pollution it is impossible to make a simple statement that a certain substance is harmful without specifying the nature and the volume of the receiving waters. In fact, some lakes and streams may contain such high quantities of particular substances that they might be called "naturally polluted." For example, in many parts of Canada there are very rich mineral deposits which give the surrounding lakes high concentrations of such elements as mercury (Pinchi Lake, British Columbia), arsenic (Red Lake, Northwest Territories), copper and zinc (Flin Flon, Manitoba). It has even been suggested that the quantities of these elements in fish flesh may be useful to prospectors who are seeking new mineral deposits.

In these circumstances, it becomes difficult to frame adequate guidelines for pollution control. At one extreme it is obviously undesirable to allow every lake and stream to degenerate to the lowest common naturally occurring denominator. This would lead to a country full of lakes and streams that combined the innate phenomena that were least conducive to living organisms. At the other extreme, if every effluent must conform to a rigid standard, it would be necessary for some operations to discharge a better quality of water than they take in. Evidently, the more judicious procedure is to ensure that the output is not of sufficiently bad quality that it creates harmful effects where it is discharged.

There remains the problem that the discharge from one source, although not in itself harmful, when added to an already polluted river may be the "straw that breaks the camel's back." This situation has cropped up in many instances in the United Kingdom and the United States, and it of course implies that the first who discharges wastes into streams may claim a right to pollute whereas the latecomers may groan that they have to comply with regulations that others do not have to follow. To anticipate this situation, all operations should be required to meet minimum standards at the point of discharge; where necessary, the criteria should be subject to revision.

In the chapters which follow there is detailed discussion of a variety of different kinds of pollution problems, particularly those that have cropped up in Canada which illustrate these general principles. They serve to underline our need for a great deal more understanding of our natural lakes and streams, and of the impact of various kinds of pollution, if we are to prevent inadvertent damage and if we are to capitalize on the opportunities for using waste discharge to ecological advantage.

chapter four

ORGANIC
POLLUTION

The major effects of pollution by organic matter are the depletion of oxygen in the receiving waters and the enrichment of
production. Whether these are a source for concern largely
depends on how much organic matter is being added to how
much water. For example, when a small amount of organic
matter of almost any kind is added to a stream, the usual result
is to slightly increase productivity. Little chips of wood, leaves,
small amounts of animal excrement, kitchen wastes, and so on
are all likely to provide just a bit more fuel for the stream
ecosystem. When the additions become substantial, they commonly cause a "plume" of changed conditions which extends
downstream for some distance before the natural state is restored (figure 14). Just below the source of pollution there
are large quantities of dissolved and suspended solids, low oxygen concentrations, and heavy growth of sewage "fungi"
(*Sphaerotilus,* actually a chain bacterium). This is a zone of
degradation and active decomposition, and is habitable only by

a few species of larger organisms such as tubificid worms. Further downstream there is a gradual recovery: oxygen concentrations rise, various species of filamentous algae of genera such as *Cladophora* develop thick green slimy mats over the bottom and at the edge. Chironomid larvae are the dominant species in a larger community of bottom organisms. This region is highly productive and it assimilates most of the nutrients that were added by the sewage into a biological system. Still further downstream there is gradual recovery to natural conditions and, as the plume of greater production becomes progressively less noticeable (figure 14), more of the natural stream organisms are present.

The greater the load of sewage wastes in a stream, the further downstream are the results noticeable. If the charge is light, then there may be only a few minor effects at the point of emission. With a heavy burden, there are clearly apparent zones of degradation, decomposition, and recovery. Extreme loadings can lead to the development in a river of an almost complete oxygen lack which can impede fish migration as effectively as a high dam. Situations of this kind are most common on small streams; they usually attract considerable attention; and they are usually quickly rectified by some type of treatment of the waste. On larger rivers, public health hazards are usually noted long before there are obviously visible signs of biological effects. For example, some of the channels of the Fraser River, near its mouth, and parts of the St. Lawrence River downstream from Montreal are not considered safe for swimming, though they evidently are not affected yet to such a degree that there is major ecological change. Nevertheless, the quantity of wastes that they carry is increasingly obvious and there are many signs of decreasing water quality and incipient major pollution.

Pollution with most organic wastes has few permanent effects. As soon as the discharge is stopped, there is a rapid return to natural conditions that reflect the normal level of inflow of organic matter. In some instances, the pollution-control measures can take the form of requiring that wastes be disposed of on land where their fertilizing effect can be put to valuable use. Spray irrigation with a dilute solution of wastes is a typical answer to these problems, and has been widely used in conjunction with meat- and vegetable-packing plants. The City of Vernon in British Columbia uses sewage wastes in spray irrigation for hay crops.

More commonly, the control of organic pollution involves large-scale treatment that in effect moves the degradation and decomposition processes out of a stream into a contained environment. Sewage-treatment plants are the prime examples. There are several methods that differ largely in the degree to which they remove substances contained in the waste they receive (figure 15).

In *primary treatment,* the sewage is first passed through bar racks or screens to remove the coarse solids. Alternatively, the coarse solids may be ground up. Grit chambers settle out sand, gravel, eggshells, coffee grounds, and the like, and then the sewage is deposited in settling basins. In these basins a sludge accumulates at the bottom and a scum on top, and both are removed. With these simple processes of mechanical sedimentation and flotation, the oxygen demand and suspended solids are cut to about one-half of their levels in raw sewage. Before it is discharged, the effluent is chlorinated to kill bacteria and other organisms that may be health hazards.

Surprising as it may seem, even this simple level of treatment is not universal in Canada. Many of the major cities on the seashore rely on the action of tides and currents to carry their completely untreated sewage wastes offshore and to diffuse them out of sight and out of mind. Victoria and Prince Rupert in British Columbia, Hull, Quebec City, and most of Montreal in Quebec, and Halifax in Nova Scotia do not treat their sewage. Vancouver has a sewage-treatment plant, but it is pretty well bypassed when there is heavy rain (a not unusual event in that part of Canada) and the plant's capacity to handle the flow is exceeded. Although the ecological consequences of this pollution have not yet proved to be severe, they all comprise potential sources of public-health hazard and ecological trouble. Needless to say, these problems can be much more serious when inland cities and towns dump raw sewage into rivers or lakes.

Secondary treatment may involve several different kinds of processes. Most commonly, it exposes the waste after primary treatment to organisms such as bacteria, fungi, and protozoa that convert the finely divided and dissolved organic matter into solids that will settle into sedimentation tanks. The two most common ways of doing this are the "activated-sludge" treatment and the "trickling filter." The former churns the primary-treatment waste to provide aeration while the organisms assimilate the organic matter. Bubbling pure oxygen through the system gives enhanced results. The resulting sludge of flocculated particles accumulates in secondary settling tanks, and the residual waste water may be recycled to either the primary or secondary processes.

The trickling-filter process provides a bed of crushed rock from 3 to 8 feet deep on the large exposed surface of which microorganisms develop a vigorous growth. Sewage is sprinkled over the bed and, while it trickles through, it is oxidized and degraded. The drainage then goes to secondary settling tanks. This method removes 75 to 85 per cent of the organic matter, and may also be followed by chlorination before discharge of the final effluent.

Both primary and secondary treatment may involve various degrees of processing the solids that have been removed from the sewage so that they may be used for land fill or for fertilizer. The common methods are anaerobic digestion, incineration, or wet oxidation.

In recent decades sewage treatment has become more complex because of the presence of many new substances in domestic waste. For example, salts, acids, certain dyes, some insecticides and herbicides may be as harmful pollutants when they leave a sewage plant as when they entered it. To solve these problems, *tertiary-treatment* procedures have been developed, including chemical precipitation of some nutrients (phosphorous, especially) and harmful substances, filtration, removal of ammonia by aeration, and adsorption by carbon columns. Many techniques are now being tested in pilot-plant operations.

A particular problem in tertiary treatment has been created by phosphates, which are contained in most detergents. The earlier detergents contained a base of alkyl benzene sulphonate, which is very stable and resulted in white foam on many lakes and streams. This predicament was relieved by the use of a linear alkylate sulphonate base, which is decomposable by bacteria (hence biodegradable). But this change did nothing to offset the enriching effect of phosphate on the growth of algae in the streams and lakes receiving the effluent. The removal of

phosphates is therefore a particularly pressing problem in tertiary treatment. The development of non-phosphate detergents is
another possible solution, and among those being considered is
the substance known as NTA (nitrylo-tri-acetate). Because it
contains nitrogen it is a potential fertilizer. Additionally, it may
have direct toxic or perhaps carcinogenic effects on fish and
other organisms. These possibilities are all being widely re-
searched at the present time in many Canadian and United
States laboratories.

It seems likely that, as the number of new compounds make
our domestic technology more efficient, it will be necessary to
make our sewage technology equal to the task of putting back
into our streams water that is as good as that which is taken out
of them. Many of the effects may be subtle. The importance of
adequate investigation prior to the wholesale use of new substances cannot be too highly stressed, as a precaution against
inadvertent pollution and against public hysteria. For example,
it has recently been speculated that the widespread use of birth-
control pills may so increase the concentrations of female hor-
mones in water as to affect fish reproduction and, for that matter, perhaps human reproduction as well, when the water is
"recycled" through the cities downstream. Fortunately, these
fears appear groundless because the steroid substances concerned are readily biodegradable, and the concentrations involved are low. Nevertheless, the "pill" affair illustrates that
there can be potentially disastrous secondary consequences of
something we may consider as a modern technological blessing.
It is accordingly necessary to carefully assess such possibilities
before there is widespread use, and to develop as further insur-
ance the most comprehensive and thorough techniques of tertiary treatment.

Problems of organic pollution are at present gaining in sever-
ity in Canada as a whole, and in the more densely populated
areas there are many examples of serious local situations which
are health hazards and incipient ecological disasters. Virtually
the whole of the St. Lawrence River system is approaching a
critical stage in which the combined effects of organic and industrial pollution will soon be grossly evident. Many small watersheds are already receiving heavy loads of sewage wastes. The
Canadian Federation of Mayors and Municipalities sent a ques-

tionnaire in 1970 to 60 municipalities, of which 47 replied. Of these, only 26 said that they had any sewage treatment, so that it would appear that close to half of our waste-water collection systems end up dumping their effluent in its raw form. Eight of the 26 gave only primary treatment. Standards for the eventual discharge vary greatly, some requiring that the effluent should place a low demand on the oxygen capacity of the receiving waters (a biochemical oxygen demand of 15 ppm) whereas others will permit a discharge that is almost ten times more demanding (120 ppm). Some sewage plants tolerate only 15 ppm of grease in the discharge, but others will release 250 ppm. In general, then, sewage treatment in Canada is primitive and variable. There are only a few model sewage-treatment plants that reflect credit on their municipalities. It should be insisted that higher standards should soon be the common, rather than the uncommon, practice.

Sewage pollution from septic tanks in rural Canada may continue to be a problem for many of our countryside lakes and streams. Leakages from septic-tank fields are rich in nutrients and, taken singly, are only small contributors that enrich the natural system. But, when cottages line a lake shore or a canal, the sum of a large number of septic-tank emissions can have effects on lake or stream metabolism. There are three obvious answers to the problem. The first, sewer systems, seems prohibitively expensive. The second, requiring that cottages and their septic fields be placed well back from the water's edge, seems unrealistic, but it takes advantage of the capacity of soil to absorb much of the nutrient materials. The third, that attention be given to developing different technologies of sanitation, would seem to be a path that will be followed in the future.

An organic-pollution problem of considerable dimensions may result from excessive discharges of animal manure where chickens, turkeys, or pigs are concentrated into small areas, or where large numbers of cattle are held in confined stockyards. This situation is of increasing concern in the prairie provinces, and as yet there have been few steps to find solutions.

Organic pollution can also occur when the various wastes of logging practices are allowed to accumulate in lakes and streams. Chips of wood and bark, small branches, and leaves and needles may all find their way into small streams, and because they are often only slowly degraded they may smother stream organisms. The erstwhile practice of building sawmills near lake shores can produce thick layers of sawdust that convert the lake bottom into a virtual desert. "Log driving" on streams usually results in much abrasion, and large quantities of bark become lodged in the back eddies of the margins. The bark is decomposed only slowly, and blankets of bark and chips may smother out a large proportion of the parts of a watercourse that are most productive of food organisms for fish. Some of these effects may eventually be passed far downstream and may ultimately be obvious in lakes that receive the drainage. There are many examples throughout the boreal forest zone of Canada, and anyone who has walked in the Canadian woods will know how common it is to find a lake or stream that bears the scars of a logging operation long after the loggers have moved on.

There is a particular moral to be drawn from most instances of organic pollution. All organic matter is engaged in an endless natural recycling. There is nothing we need do but take advantage of these processes to ensure that organic pollution does not occur. Pollution ensues when we concentrate organic material and release discharges that exceed natural assimilative capacities. What we need to do, then, is to redistribute our organic wastes, returning them to their sources. In the long run, to sustain our capacity to produce, we should ensure that we place as much emphasis on dispersing our wastes as we do on gathering food and fibre.

chapter five

TOXICANTS

GROSS POLLUTANTS

Almost anything added to water in sufficient amounts will be
toxic to fish and other aquatic organisms. Even common salt,
when dumped into a lake or stream in large quantities, will
quickly eliminate almost all life. This is to be expected, because
it is tantamount to artificially creating a saline environment in
which most creatures are unable to maintain an osmotic balance. In effect, the salt extracts the water from living tissues and
they become literally "pickled." Similarly, large quantities of
acid or alkali of any kind will so change the pH of water that
fish and other organisms are killed by their tissues being burned,
or by their inability to maintain normal concentrations of ions
in their body fluids against very high gradients of pH. Many
industrial effluents produce gross effects of these kinds and are

thus toxicants by virtue of their very high concentration. Additionally, many of them are complex mixtures of many substances, and it is seldom easy to pinpoint what it was in the mix that did the damage.

The effect of particular gross contaminants largely depends on the chemistry of the receiving water. If, for example, a lake or stream is quite alkaline and has a high natural dissolved mineral content (e.g., almost any lake in the Prairies), it has a large buffering capacity and the pH is not changed except with large pollution loads. By contrast, when the natural waters are low in mineral content (as almost any lake in the Canadian Shield), even small amounts of acid can markedly change the pH. In general, fish can tolerate a range in pH from 6.5 to 8.5. If the addition of a pollutant raises or lowers the pH beyond these limits, then fish either are killed or move away. A particularly vivid example of a pH change is the effect of sulphur dioxide emission in smoke from the large smelting operations at Sudbury. The sulphur dioxide combines with water to become dilute sulphuric acid, and for a radius of 50 or more miles around Sudbury an acid rain has gradually changed the acidity of many of the surrounding lakes. Because the waters of the lakes are very soft and contain low amounts of dissolved minerals, only small quantities of acid have a drastic impact. In more than 30 lakes fish have been completely exterminated, and in at least as many more some species have disappeared and the others will presumably be gone soon. The same result would not be expected with a similar discharge of sulphur dioxide in a region where lake waters contained more dissolved mineral salts —on the prairies for example.

Another similar example of gross pollution is the acid waste that is created in the drainage from some kinds of coal mines. Throughout the whole of eastern North America, drainage from strip coal mines is highly acidic. With a substantial rain storm, the many small puddles of very acid water are swept into streams and cause an immediate kill of fish and other organisms. When this acid condition is combined with sewage discharges the result can be particularly unpleasant because the decomposition that usually "processes" sewage is arrested. The usual result is the development of extensive growths of slime bacteria— forming an environment hostile to most aquatic organisms and certainly unpleasant to look at (and to smell). In many parts of the Atlantic provinces of Canada, acid mine wastes are a major pollution problem, but they apparently are not a serious concern in the coal-mining operations of western Canada.

In general, gross pollutants are not a particularly serious concern in Canada because we do not as yet have heavy concentrations of manufacturing industries, but on a small local scale they can be drastic in their effects. This is particularly so because, typically, the discharges can come in short heavy doses when there is an accident, or a plant "clean up," or an intermittent purge of "used up" materials. In a space of a few hours (or even minutes), a sharp strong surge of a substance that is nontoxic at low concentrations may wipe out the organisms in a section of a stream. These cases of pollution are often not discovered until some time after, when it is frequently difficult to trace the source. The obvious cure is good housekeeping practices in industrial operations. In many instances the discharged wastes can be economically recycled or used in other ways. It is important that government agencies know what types of discharges are likely to occur from any industrial process so that preventive measures may be enforced. At present, most government agencies in Canada work only at scolding offenders after a pollution accident has occurred, whereas, ideally, their work, like that of firemen, should be largely devoted to prevention. When new plants are being constructed, it is common practice for pollution-control agencies or government departments to give advice on what pollution-abatement devices or procedures should be installed. In the best situations, a "permit" will not be issued unless this is done. But for existing plants there is no law that calls for closure if proper equipment and procedures are not installed, only provision for a fine if a pollution incident occurs and is detected, and if the case is taken to court successfully.

HEAVY METALS AND MINING

Heavy metals are highly toxic to fish and are, of course, common pollutants from the mining-industry operations which extract them, as well as from the various industries which use them. The lethal level of heavy metals is particularly dependent on the concentrations of other salts in the water. The effect of some heavy metals is to precipitate the proteins in the mucous that covers the gills of fish, thus inhibiting respiration. The same can happen in the mucous on the skin, plugging the olfactory organs and preventing the fish from detecting the presence of toxic materials in the water.

It is characteristic of many mining operations that the ore is concentrated by grinding it into a fine flour which is then treated with a variety of chemicals to extract the valuable metals. The waste tailings are then discarded. Where they are discharged directly into a stream, they quickly kill the organisms on the bottom by plugging their gills and abrading their exposed surfaces. The tailings also fill the spaces between the gravel particles, making the stream bottom like a paved surface and rendering it useless for fish spawning. In these circumstances, the effect of the toxic metals and cyanides is almost incidental. Where the tailings are impounded, which is now common practice in Canada, they settle out in ponds. If there is seepage from the ponds, it may be highly toxic.

Where tailings are discharged into lakes, they may damage bottom organisms as they settle and, because the finer particles settle slowly, there is usually some adverse effect on plankton production. The concentrations of heavy metals are rapidly diluted in the large volume of lake water, but their presence is eventually reflected in increased amounts of metals in the fish tissues, particularly in the liver where they accumulate. It is not known what the chronic effects of these sub-lethal doses may be. In many natural waters the fish contain relatively high concentrations of some metals, and it has been suggested that fish livers be used by prospectors to indicate drainage areas that may contain ore bodies. The naturally occurring situation is highly variable and underlines the fact that single standards for discharges into lakes of wastes containing heavy metals are virtually undefinable. In general, it is undesirable to increase heavy-metal concentrations above background levels, and standard tailings-pond disposal methods can usually accomplish this. Interestingly, in many instances it has been found that tailings may be used to refill old workings, and in some situations new technologies have made it profitable to reprocess the old tailings. Unless the mining operation is a quick-plunder venture, good pollution abatement can be made profitable.

The Canadian experience of mining pollution has been extensive. Literally hundreds of small mines have, from time to time, dumped tailings and toxic wastes into the handiest streams. In more settled regions, a steady social pressure has gradually brought the offenders under control, and tailings ponds are the rule now rather than the exception. A number of streams that were once badly polluted have at least apparently returned to natural production (the Similkameen River in British Columbia, for example). One of the problems that remains is that many of the tailings deposits have been abandoned and persist as ugly patches of finely ground rock that do not support vegetation. Covering and planting of tailings piles should be made a mandatory requirement for all "completed" mining operations (figure 16).

Pollution from mines may continue to be a problem in the hinterlands of Canada for many years yet. In the more remote areas, the number of people who use the streams and lakes is small in relation to the numbers who want industry and feel that mining would be uneconomic if anti-pollution measures were taken. Almost invariably, this kind of reasoning has proved to be near-sighted. The hinterland of today will be increasingly used as a playground tomorrow. Mining developments in northern Canada should conform to the same standards of pollution control as those that southern mines should have developed thirty years ago.

MERCURY POLLUTION

A particular heavy-metal pollution problem has recently come to the fore in the form of high mercury concentrations in fish flesh. The contamination of lakes and streams by mercury had been well documented in Sweden in the mid-1960s, but it was not until September 1969 that it was recorded as a problem in Canada. In the course of studies of possible mercury pollution from the pulp-mill operation at Prince Albert, Saskatchewan, a graduate student (now Dr. Wobeser) examined the flesh of fish in water receiving the pulp-mill effluent. It was logical to wonder what mercury levels might be in a stream that did not contain pulp-mill wastes, and for this purpose Dr. Wobeser was provided with specimens from the South Saskatchewan River near Saskatoon by Dr. M. Atton of the Saskatchewan Research Council. One of these specimens was a pike, the flesh of which contained 11 ppm of mercury—a level over 20 times that considered "safe" for human consumption (0.5 ppm). Further samples confirmed that predatory "fish-eating" fish contained high concentrations of mercury.

The Fisheries Research Board of Canada was advised of these findings and soon had confirmed that there were high mercury concentrations coming into Lake Winnipeg from the Saskatchewan River, and that, in addition, the lake was also getting mercury from a chlor-alkali plant at Dryden, Ontario. In March 1970 a study commissioned by the Canadian Wildlife Service disclosed high levels of mercury in fishes from Lake St. Clair, and again chlor-alkali plants seemed to be responsible. Shortly afterwards, high mercury levels were confirmed in the western end of Lake Erie. A large number of investigations have followed and it has been established that mercury concentrations in flesh of fish-eating fish are relatively high in many Canadian rivers that have industries or large cities on their banks (the St. Lawrence from Cornwall to Trois-Rivières; the Ottawa from Ottawa to Montreal; parts of Lake Huron, and so on).

There are many obvious sources of pollution by mercury. At many sites, chlor-alkali plants are the obvious major contributors; in manufacturing chlorine and hypochlorite for pulp-mill bleaches and other uses, they employ a mercury cathode that each day may shed many pounds of mercury (half a pound per ton of chlorine produced). This sort of loss has now been corrected at many Canadian plants of this kind, with the daily amount only a small percentage of previous levels.

Mercury compounds (particularly phenyl mercuric acetate) are used to control slime in pulp and paper operations, and where plant operations are sloppy there may be discharges into the environment. In January 1970 all Canadian mills were forbidden use of these mercurial slimicides, and this source also would seem to be under control (figure 17).

Mercury is also used in treatment of seeds for disease control: almost 70 per cent of all grain seeds in Canada in 1967. This usage seems to be declining, although it is still an important

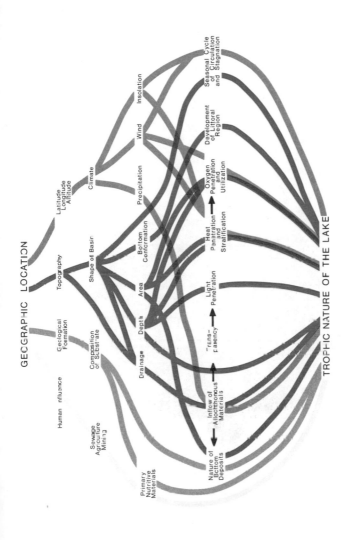

FIGURE 1. A diagrammatic representation of the various factors that influence the characteristics of lakes. Because of the large number of factors and their complex interrelations, many of which are not shown, virtually all lakes can be considered unique. This diagram was first prepared by the University of Saskatchewan's Dr. D. S. Rawson, a pioneer in the study of Canadian lakes.

FIGURE 2. Production and decomposition zones in lakes. The basic production of plant material occurs only to the limit of light penetration. The sun provides the energy for photosynthesis of aquatic plants. Decomposition of dead plants and animals takes place on the lake bottom. Rooted aquatic plants are important in lake productivity—lakes with longer and more irregular shorelines are more productive.

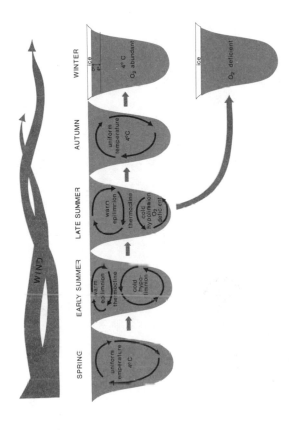

FIGURE 3. The temperature cycle in temperate lakes. In the spring and autumn, when a lake has the same temperature throughout, there is free circulation from top to bottom, recharging the surface waters with nutrients and the deeper waters with oxygen. If the autumn circulation does not occur, then the lake may be deficient in oxygen during the winter. In mid-summer, the lake has three layers: a warm upper layer (epilimnion), a cold lower layer (hypolimnion), with a zone of rapid temperature transition between (thermocline). Wind does the work of circulation that distributes the sun's heat through the upper layers of a lake.

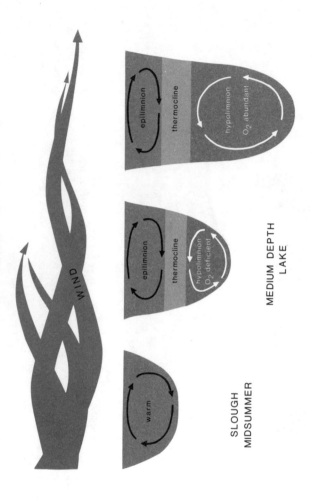

FIGURE 4. Influence of lake depth on the relative sizes of the upper and lower zones. In shallow sloughs, the wind will mix heat to the bottom so that the whole lake is warm. In lakes of moderate size, there is a relatively small lower layer which is not circulated in mid-summer. In large deep lakes the cold lower layer may constitute most of the volume and the thermocline may be thicker.

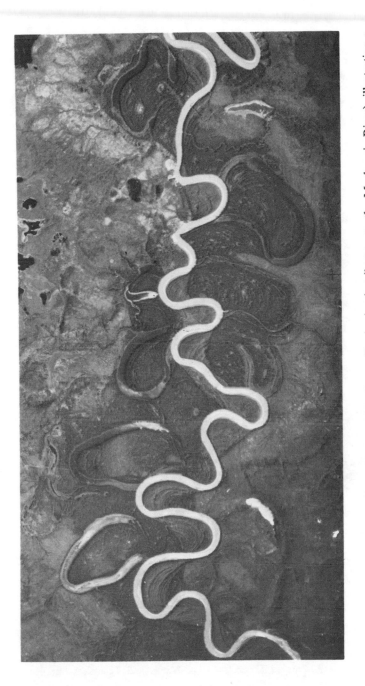

FIGURE 5. Aerial photograph of Hume River, Northwest Territories (a tributary to the Mackenzie River), illustrating a meandering stream and oxbow lakes. (Photo: Department of Energy, Mines and Resources, Ottawa.)

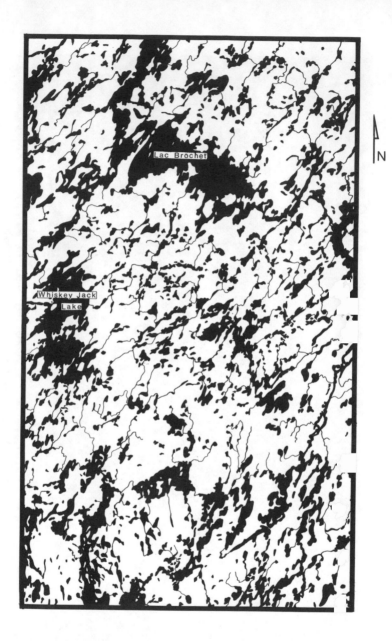

FIGURE 6. Lakes of northern Canada. Approximately 1,700 square miles of northern Manitoba, at the Saskatchewan border, illustrating the large number of lakes in northern Canada. Modified from National Topographic Map, Canada Department of Mines and Technical Surveys, Ottawa.

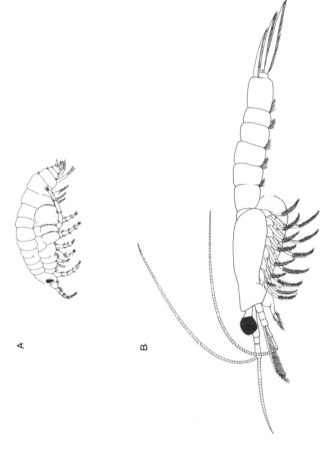

FIGURE 7. *Pontoporeia affinis* (A) and *Mysis relicta* (B). These two crustaceans are important fish-food organisms in most deep cold Canadian lakes east of the Continental Divide. They are relicts of our recent glaciation. They require cold temperature and high oxygen concentration, and are vulnerable to changes in the pollution of lakes. (A) After Ward and Whipple, *Freshwater Biology*; (B) After Bullough, *Practical Invertebrate Zoology*.

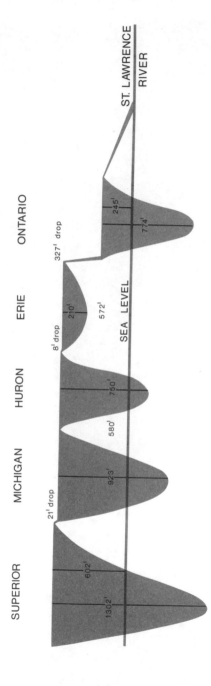

FIGURE 8. Profile of the Great Lakes indicating their depths with reference to sea level. With the exception of Lake Erie, the Great Lakes are oligotrophic in their natural state. They are unusual in that they have a relatively small drainage area in relation to their size.

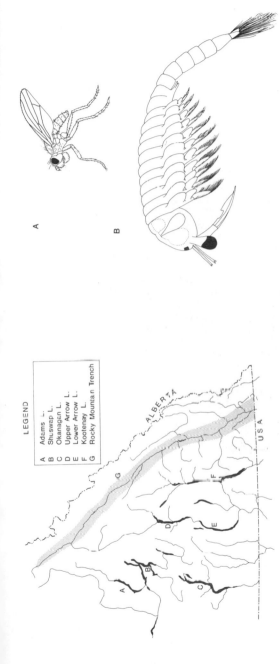

FIGURE 9. The major mountain lakes in British Columbia, all of which have had runs of salmon at some time in their history. Some have been cut off by natural obstacles (Kootenay) and others by man-made dams (Arrow Lakes). There is now a small run of sockeye salmon to Okanagan Lake. Shuswap Lake has a run of several million sockeye salmon every fourth year; most of these fish spawn in the Adams River which flows from Adams Lake to Shuswap Lake. The Rocky Mountain Trench is a major geological structure and is referred to in the last chapter of this book.

FIGURE 10. *Ephydra* (A) is a fly that lives as a larva in saline lakes. *Artemia* (B) is a brine shrimp found in saline lakes all over the world. These two organisms are among the few species of animals that can live in extremely salty lakes that may be many times more salty than the sea. (A) After Essig, 1942, *College Entomology*; (B) After Kaestner, 1970, *Invertebrate Zoology*.

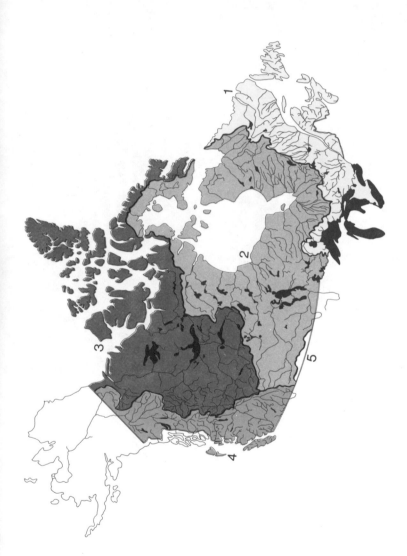

FIGURE 11. The five major drainage regions of Canada: (1) Atlantic; (2) Hudson Bay; (3) Arctic; (4) Pacific; (5) Gulf of Mexico (Mississippi).

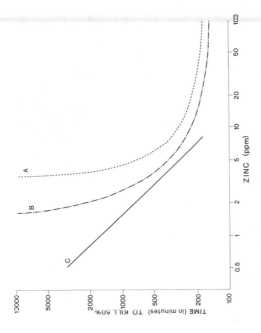

FIGURE 12. The common method of describing toxicity of a compound to an organism. At low concentrations the toxic material, copper, does not kill any of the organisms in a 96-hour exposure period. At high concentrations none of the animals survive in an exposure of 96 hours. The TLM (Tolerance Limit Median) is the level at which 50 per cent of the animals survive in a 96-hour exposure period. After Warren, 1971.

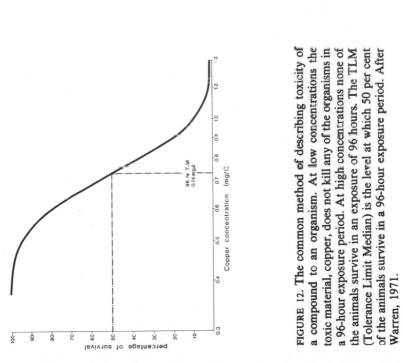

FIGURE 13. The effect of toxicity of zinc sulphate on rainbow trout in waters of different hardnesses. Curve A = pH 7.6 to 7.8; Curve B = pH 7.0 to 7.2; Curve C = pH 6.5 to 6.7. It is typical for various characteristics of water to influence toxicity, thus no single figure can be given as a "safe" level for a toxicant. After Goodman, Edwards, and Lambert, 1965.

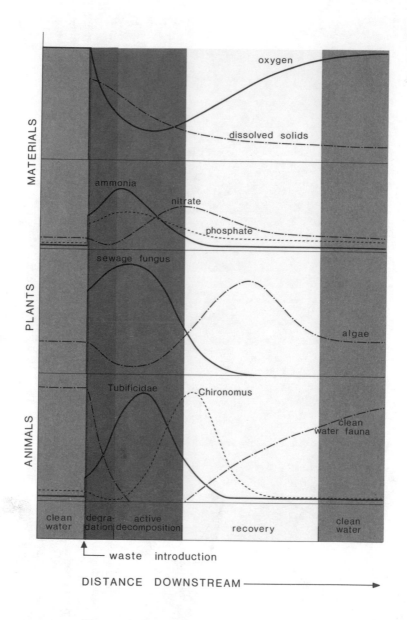

FIGURE 14. Changes in stream conditions downstream from a discharge of sewage effluent. After Hynes, from Warren, 1971.

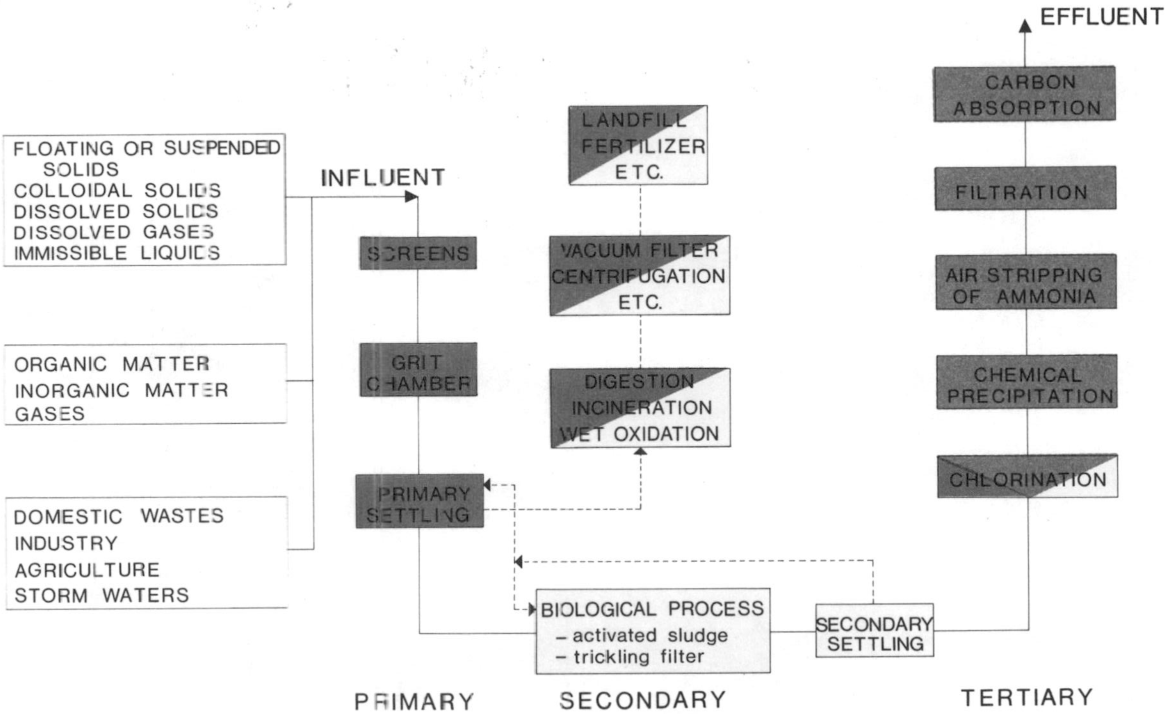

FIGURE 15. Flow-diagram representation of various kinds of sewage-treatment facilities. Some cities and towns in Canada have none at all; some use primary treatment; some use secondary; and a few are currently developing tertiary treatment.

FIGURE 16. Typical tailings pond at a mining operation (Endako, British Columbia). Tailings are the finely ground rocks from which the valuable metals have been removed. They are pumped into the pond as a "slurry" and the water then evaporates or is drained off. When the mining operation is completed, the tailings pond should be covered with soil and planted in natural vegetation. This has seldom been done in Canada in the past. (Photo: G. D. Taylor.)

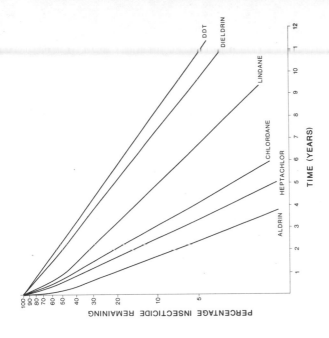

FIGURE 18. Diagram indicating the persistence of some common insecticides in the soil. After Edwards, *In:* Goodman, Edwards, and Lambert, 1965.

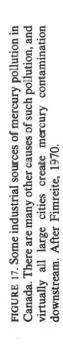

FIGURE 17. Some industrial sources of mercury pollution in Canada. There are many other causes of such pollution, and virtually all large cities create mercury contamination downstream. After Fimreite, 1970.

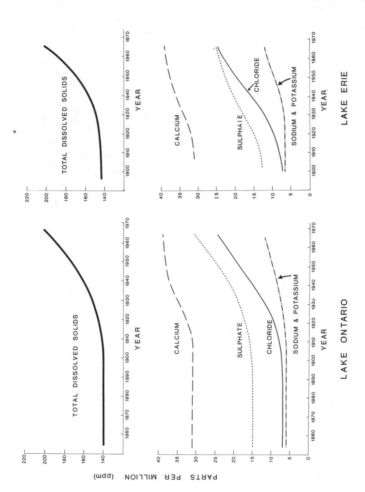

FIGURE 19. Changes in the chemical content of Lakes Erie and Ontario. The very sharp increase in the last 50 years has resulted in the eutrophication of the Great Lakes, particularly of Lake Erie. After Beeton, *In:* National Academy of Sciences, 1969.

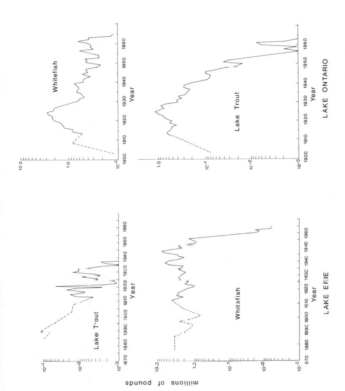

FIGURE 20. Changes in catches of lake trout and whitefish in Lakes Erie and Ontario. The declining harvest is a consequence partly of eutrophication and partly of overfishing. Many other species of fish have also been seriously affected in most of the Great Lakes. After Beeton, *In*: National Academy of Sciences, 1969.

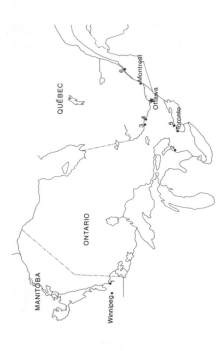

FIGURE 22. **Effect of fine sediments on the emergence of salmon from their eggs.** The newly hatched salmon are called alevins. Fine sediment in streams commonly results from logging and other activities that accelerate processes of stream erosion. From: U.S. Department of the Interior, 1970.

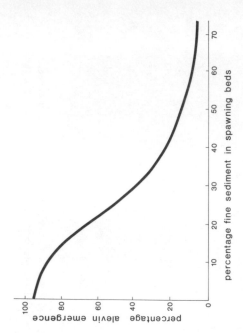

1 WHITESHELL NUCLEAR RESEARCH ESTABLISHMENT
2 BRUCE NUCLEAR POWER DEVELOPMENT
3 NUCLEAR POWER DEMONSTRATION STATION
4 CHALK RIVER NUCLEAR LABORATORIES
5 PICKERING NUCLEAR POWER STATION
6 GENTILLY NUCLEAR POWER STATION

FIGURE 21. **Atomic energy establishments in Canada.** In the future there probably will be many more nuclear plants constructed in Canada. **After Marko, *In:* Barry, 1972.**

FIGURE 23. Typical log jam on a salmon-spawning stream. This kind of situation commonly occurs as a consequence of intensive and careless logging practices. (Photo: Steve Zablosky.)

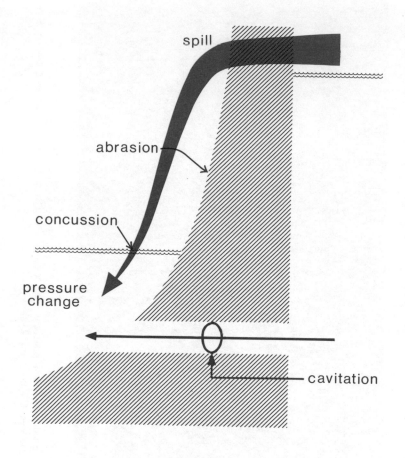

FIGURE 24. Some effects of dams on young salmon that may be killed whether they leave a reservoir by the spillway or through underwater orifices. On the spillway route, salmon may suffer abrasion on the surface or concussion when they hit the tailwater. If they go through the lower route, they may be killed by cavitation effects, by the turbine blades, or by the sharp change in pressure as they reach the tailwater.

FIGURE 25. One of the many versions of the NAWAPA scheme (North American Water and Power Alliance) for continental water diversions. Any of the components would have enormous consequences on the biology of lakes and streams. Adapted from the NAWAPA Conceptual Plan by the Ralph M. Parsons Company, Los Angeles, 1966.

(1) Rocky Mountain Trench diversion of the Yukon River and other northern British Columbia rivers to the United States.
(2) Canadian Great Lakes Waterway diverting the Peace and Saskatchewan Rivers.
(3) Dakota Canal diverting the Saskatchewan River.
(4) Hudson Bay Seaway.
(5) James Bay Seaway and Knob Lake Barge Canal diverting many Quebec rivers.
(6) Eastern aqueduct.
(7) Rio Grande aqueduct and Colorado aqueduct.
(8) South West Canal, a continuation of the Rocky Mountain Trench scheme.

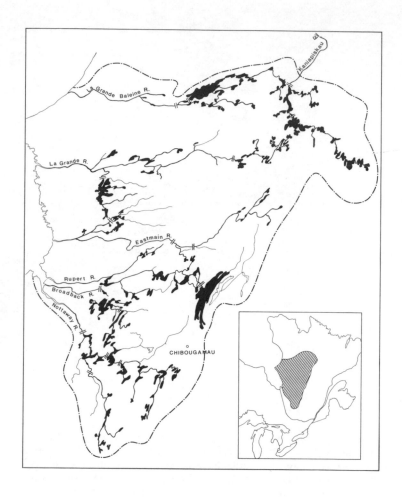

FIGURE 26. James Bay Hydroelectric Power Development Project. This scheme is now in the early stages of construction. Ultimately, it could flood 6,000 square miles of land, and would influence the drainage of about one-quarter of the area of the Province of Quebec. Adapted from *Nature Canada,* 1972.

source. Moreover, these treatments contain alkyl mercury compounds that are more acutely toxic but are eliminated less slowly than other organic mercury compounds.

Other sources of mercury are many and various and are much less easily controlled. Almost all large cities generate a downstream mercury contamination. Mercurial compounds are used in horticultural sprays, as turf fungicides, and as mildew-proofing substances. It is a common metal in heavy-duty electrical switches. Most laboratories use small quantities of mercury or mercuric salts in various experimental test procedures and in a variety of instruments. Every year, 3 million mercury thermometers are broken in Canada! The element also finds its way into water because it is released (in gaseous form) from the burning of fossil fuels such as oil or coal, and then comes down with the rain.

From these and many other small sources (dental preparations, fluorescent lights, manufacture of plastics, hats, and so on) a city of only modest size will create a mercury contamination downstream that is sufficient to induce concentrations in fish-eating fish higher than is considered safe for human consumption. Such fish as lake trout are particularly likely to have high levels of mercury because they are predominantly fish-eating. Pike and walleye are also usually relatively high in mercury content where they are exposed to such a source, and may concentrate the metal to levels more than 5000 times those in the water. Whitefish are usually much lower in mercury content, reflecting the lesser concentration in non-fish-eating species.

The many investigations that have followed the 1969 "discovery" at Saskatoon have abundantly demonstrated that mercury contamination of lakes and streams is widespread in Canada and arises from many sources. Although this impact of man on his environment is obviously undesirable, it may well be that the potential effect of mercury poisoning has been somewhat exaggerated in the public mind. Man and other animals have substantial capacities for eliminating mercury and, unless there is a continued substantial daily dose, mercury (in at least some of its forms) will not accumulate to dangerous levels. Thus, the occasional fish eater (and Canadians only consume on the average about 10 pounds of *all* kinds of fish per year so most of them

must be occasional fish eaters) is probably in little danger of
mercury poisoning from eating contaminated fish. Those that
delight in eating fish and are really big consumers should obviously consider where their fish is coming from.

The effect on fish and other organisms of a steady diet of
mercury compounds may be quite profound and, unfortunately,
the mercury that has been discharged into our lakes and streams
may persist for a long time. Much research is being done on
mercury contamination at the present time. The metal may be
discharged in a variety of forms: as elemental divalent mercury,
metallic mercury, phenyl mercury, methyl mercury, or as alk-
oxi-alkyl mercury. In the natural environment, all forms can
undergo many transformations, the one with most serious
consequences being methylation of inorganic mercury. Once
methylated (probably by bacteria in mud), mercury is readily
absorbed by organisms, and is only slowly excreted. For example, fish may concentrate methyl mercury up to 5,000 times by
absorption through the gills and by eating phytoplankton. In
addition, fish may convert inorganic mercury into methyl mercury, the change occurring in the mucous of their skin or
through the agency of intestinal bacteria. After it is absorbed,
mercury tends to localize in the the kidney and liver, from
which it is excreted; but, with continued accumulation, it is
found in increasing concentration in the brain and muscles.

A number of solutions to the mercury problem have been
suggested. Dredging of the mud from heavily polluted areas has
been tried, but because of the stirring and mixing that occurs it
has not been successful. Covering the polluted sediments is also
being tested. The bacteria responsible require only a little oxygen to do the methylation, but this small amount can be cut off
with a layer of 2–3 cm of sediment. More extensive sewage
treatment can diminish the supply of organic matter that supports the large bacterial populations and thus slow the methylation process. Hydrogen sulphide can be used to precipitate mercury, but is itself toxic and as a short-term cure would have
drastic effects.

Much more research is needed if we are to devise methods of
undoing the damage that has already been done to Canadian
lakes and streams from mercury pollution. Large accumulations
of this element will continue to recycle in the aquatic food

chains and to contaminate freshwater fish, unless ways are found of putting the mercury out of biological circulation. Further research may also disclose that other heavy metals have become semi-permanent environmental contaminants. Investigations along these lines should be pressed, because the sooner we are aware of potential hazards, the sooner something can be done to rectify them.

OIL POLLUTION

Oil is a name that is given to a great variety of compounds, each of which is, in itself, a fairly complex mixture of hydrocarbons. Oil of one kind or another can enter fresh water from a wide variety of sources, each with its own particular characteristics. At one extreme, for instance, pollution may be caused by leakages of diesel fuel oil which is made up of relatively volatile hydrocarbons. At the other extreme, waste from industry may be heavy oil and grease, containing a large quantity of impurities, and may enter the water as a thick viscous sludge. On the average, what we usually mean by "oil" is a relatively stable hydrocarbon material which is slow to degenerate and which may persist for long periods of time; it is perhaps one of the most insidious and serious sources of damage to aquatic organisms.

Oils and the emulsions that are commonly used to disperse them have an immediate substantial effect on fish. They generally stick to the gills and the fish can only get rid of the oil by producing large quantities of mucous. This leads to respiratory distress and, occasionally, to asphyxiation. In addition to these surface irritant effects, most oils contain a water-soluble fraction that is directly toxic to fish.

The effects of oil on other aquatic organisms are also profound. Many aquatic insects and crustaceans come to the surface to breathe, and oil floating on water not only cuts off the source of oxygen but also has toxic effects directly on these organisms. The heavier components of most oils commonly form a sludge on the bottoms of lakes and streams. Because it is viscous and stable, this sludge makes an effective blanket that blocks oxygen transfers and directly poisons the bottom organisms. The blanket is, of course, long lasting. Even the continued accumulation of oil from heavy use of outboard motors should give cause for concern in many of our recreational lakeshore areas.

Oil also has disastrous effects on water birds, and there are many well-documented instances in Canada where large numbers have been killed by oil pollution. The oil destroys the insulating properties of the feathers; birds that have been exposed to oil can quickly die from exposure. When they clean themselves, wiping their feathers with their bills, birds ingest oil and may then suffer from its direct toxic action.

It is not generally realized that oil compounds are toxic to most vertebrates, so that the presence of oil in a water supply can constitute a serious human health hazard. In brief, oil is a major and serious pollutant, and its widespread use in a great many human activities poses a large number of problems of pollution control.

The problem of what to do with waste oil products has bothered engineers for many years. Recently, however, it has been discovered that treatment facilities can be devised which rely on bacterial decomposition of oily materials. The development of these centres in association with sewage-treatment plants, oil refineries, and many kinds of industries is much to be encouraged.

Serious consequences can follow the large-scale spill of oil from a pipe line. As everyone knows, the possibility of a pipe line traversing the Mackenzie basin to bring arctic oil to the more populated parts of Canada has received extensive recent study. It is obviously important that any pipe line should incorporate a system of pressure-activated valves which close as soon as there is a break in the line. In this way the spill is limited to only the amount of oil that is between two of the valves. However, even this arrangement only ensures that the quantity of oil dispersed will be large instead of enormous.

Although a substantial amount is known about the effects of oil on organisms, most of the research has been done in relatively warm climates. Therefore, it is a matter of considerable conjecture to state what the effects of a large oil spill would be in the Arctic or Subarctic—especially in the winter. It might be expected that the oil would be even more stable at these temperatures and would take much longer to be degraded by biological action. Almost certainly, a substantial oil spill would be associated with a significant loss of bottom organisms and fish, and the recovery would probably be extremely slow.

Recently there has been discussion of the possibilities of constructing an eastern arctic oil pipe line down the coast of Hudson Bay. Again, this can only be looked on with grave concern.

The alternatives to these long pipe lines that might spill oil into fresh water are perhaps of even greater concern. The transportation of arctic oil by ship has been widely considered as one option. If only one of these large oil tankers were to sink, the resulting oil spill could have far more devastating effects on more organisms in the sea than the rupture of an oil pipe line would have on organisms in fresh water. Large-scale oil transport by either method would seem to give us a Hobson's Choice: the dangers of oil pollution are serious in both.

HERBICIDES AND PESTICIDES

Most modern pesticides are a product of the enormous development in organic chemistry in the past century. Literally hundreds of thousands of organic compounds have now been synthesized. Probably close to a hundred thousand organic compounds are known that have a potential use as pesticides. At least ten thousand have been, or are being, produced commercially. This is obviously such a tremendous variety of compounds that the literature on them and their effects on organisms is of staggering proportions. Even a simple listing of known pesticides and their toxicities would occupy thousands of pages. Of necessity, then, this discussion will be concerned only briefly with some of the major pesticides and some of their effects on freshwater organisms.

In the great majority of instances, pesticides have been developed to deal with insect pests. Insects are arthropods and belong to the same phylum (major animal group) as many freshwater plankton and bottom animals. Many of the most important aquatic fish-food organisms are insects. It is scarcely surprising, then, that pesticides can have a devastating effect on the many small animals that are key components of aquatic ecosystems. To be sure, most dispersions of pesticides into fresh water are inadvertent. Sometimes the amount of the chemical spread on the land is more than is necessary and some of the excess washes off into streams. In some instances, aerial sprays are applied when there is a breeze and the pesticide is blown into the nearby lakes and streams. All too frequently, pesticide containers are washed out in the handiest creek. These "accidents" would be of relatively small concern, except for the fact that many pesticides are extraordinarily toxic to aquatic organisms.

In addition to these shotgun effects of pesticides, many of them are directly toxic to fish and, for that matter, to man as well. For example, the organo-phosphorous insecticides influence the transmission of nerve impulses by blocking the action of the enzyme choline esterase. In effect, they throw the switches at the nerve endings so that they remain in the "on" position: the result is a complete breakdown of nervous coordination. This mechanism is common to a wide range of animals. These compounds also gain entry to an organism simply. They can penetrate through the skin, through the digestive tract, or through the respiratory system. The fact that they are so potent and so readily taken up has earned them the name "nerve gases."

Other insecticides have lethal effects on fish but are not so dangerous to man and other vertebrates. For example, the substance known as rotenone is a very useful insecticide and is also potent in killing fish. It was used by primitive tribes in South America to help them in collecting fish to eat. It is also commonly used to eradicate a fish population which is considered undesirable. The insecticide toxaphene is notorious for its toxicity to fish and has also been used for controlling fish populations. Even doses as low as one-half part per *billion* are lethal when persisting over a long period of time. Moreover, these low concentrations also kill many plankton and bottom organisms. At Paul Lake in British Columbia, toxaphene was used to eradicate redside shiners that had reduced production of Kamloops trout. The toxaphene "eliminated" the shiners, but it also wiped out the amphipod, *Gammarus,* one of the major food items of trout, along with most of the aquatic insects, and several of the more important planktonic food organisms. The plankton in the lake did not return to its natural condition until five years after the toxaphene treatment. *Gammarus* were reintroduced into the lake, but even eight years after the treatment they were present only in small numbers on about one-half of the shore areas of the lake. Although this was a drastic and deliberate addition of a potent insecticide, it demonstrates the tremendous impact that some modern chemicals can have on aquatic systems.

A large number of insecticides are toxic both to fish and to a wide variety of aquatic organisms. The effects of indiscriminate use of insecticides can be disastrous to aquatic communities, and in many instances serious local pollution has been reported.

There may be many situations in which insecticides have had effects that were attributed to other causes. In addition to their lethal action, insecticides may have sub-lethal effects—they influence behaviour and impair normal physiological processes. They reduce growth rate; they increase the susceptibility of fish to infection. They probably have similar effects on other aquatic organisms. For instance, it is well known that DDT inhibits moulting in aquatic insects. It also influences the ability of young salmon to learn and to demonstrate some basic reflexes.

One of the most serious insecticide problems that has developed in recent years arises from the chemical stability of that group of insecticides known as halogenated hydrocarbons, including the famous compound DDT. It is quite reasonable to say that the planet Earth is now seriously contaminated with DDT (and its derivatives, such as DDE) and other halogenated hydrocarbons. Because they are stable, these compounds may persist in the soil for many years after they are applied (figure 18). They may also accumulate in organisms that take in less than a lethal dose. DDT is generally stored in fatty tissue and thus fish which have eaten a large number of insects that died from DDT may contain as much as 2,500 ppm in their flesh.

It is not well known what the effect of this accumulation is on the animal. It may become serious only if the animal has to live off its fat for some period of time. However, there is one important side effect. Fish eggs contain a large amount of oil. If a fish is carrying a large dose of DDT, then its eggs will contain large quantities, and these eggs show a much reduced fertility. One of the disappointments of the introduction of Pacific salmon into the Great Lakes was to find that their eggs contained so much DDT they they were largely infertile. For perhaps somewhat different reasons, DDT also has effects on successful reproduction in birds: it causes the shells of the eggs to be thinner and hence more easily broken. In addition, it apparently influences the normal nesting behaviour of the parent birds.

For these several reasons, DDT has emerged as a major problem to those concerned with conservation of natural aquatic communities in most parts of the world. It may prove that DDT does not always accumulate in stream situations. For example, in a study of the North and South Saskatchewan Rivers, there were only small residues in fish after twenty years' use of DDT to control blackfly larvae. Accumulations in lakes may be much longer lasting. Substantial residues have been reported, especially for predaceous fishes and fish-eating birds, in many Canadian lakes, including the Great Lakes.

The foregoing is only a brief sketch of some of the effects of insecticides on aquatic communities. The fact that there are literally thousands of organic compounds whose aftermaths have not been studied is not reassuring. It is astonishing, when one reflects on it, that we continue to allow the use of compounds whose actions we may easily discover some day soon to be serious, or even disastrous. It is a special kind of Russian Roulette, with the pistol pointed at our children's heads, if not at our own.

In general, substances used as herbicides are less toxic to fish and aquatic organisms than are insecticides. For example, several of the common commercially sold herbicides are not lethal to fish even in concentrations of 1,000 ppm. Unfortunately, though, some of the most widely used herbicides, such as 2,4,5-T (trichlorophenoxy ethanol), will kill fish at 1 to 2 ppm, and others are even more toxic. The careless use of these herbicides can be particularly harmful to such fish as young salmon and trout in small streams.

Herbicides are frequently used to control aquatic vegetation when it becomes a nuisance near swimming areas, or when it fouls irrigation canals. One of the most common is copper sulphate, which is particularly effective for controlling algae; it is also used to kill snails that carry the small organisms (larval stages of Trematode worms) that cause swimmer's itch. At concentrations of less than 1 ppm copper sulphate is generally harmless to fish, but it may be difficult to apply it in a uniform way. Most other herbicides used for controlling aquatic vegetation are also relatively safe, although, again, those that contain 2,4,5-T can be highly toxic.

Where large growths of aquatic plants are killed by herbicides, there may be a short period when the decomposition of the plants reduces the oxygen concentration in the water, with possible effects on fish and other organisms. These sorts of problems are most common in small ponds and lakes, and are not of major concern in most large bodies of water in Canada.

Considering how dangerous they are, both to people and to wildlife of all kinds, it is surprising that insecticides and herbicides are so easy to buy and to spread. Almost any hardware store in Canada dispenses a wide variety of pesticides but virtually no literature or advice on the dangers of their use. As a consequence, they are used improperly, usually to excess, and with little appreciation of their eventual impact. The villain, in the minds of most, is DDT, and many Canadians threw away their stock of this product when they came to know its effects. The result, of course, was a large short-term dose of DDT on garbage piles and in sewage outfalls! Much of the danger of wide use of DDT is now realized, but the same appreciation has not extended to many of the other items on the shelves of every hardware. It should be underlined that most pesticides should be handled only by competently trained workers who specialize in knowing the hazards. That's what we do with dynamite, and in many ways some of our common pesticides are just about as dangerous.

PULP-MILL WASTES

The pollution problem associated with pulp mills is not only that their wastes are toxic, but that they produce tremendous volumes of waste. A modern pulp mill takes in 30,000 gallons of water for each ton of pulp that is produced. This water is eventually discharged and, with it, large quantities of waste chemicals, undigested components of wood (or whatever material is being used for making the pulp), and a complex mixture of organic compounds. In a day, the total discharge from a large mill may be 50 million gallons. The total quantity of organic material is so large that there is a heavy oxygen demand on the receiving waters. In addition, there is continuous production of a variety of compounds that are highly toxic to fish. A pulp mill thus combines many of the features of organic pollution with the toxic pollution that is usually associated with industrial processes.

There is a wide variety of kinds of pulp-mill operation, and the wastes are not the same from the different methods. In the simple ground-wood process, which is used for producing low-grade pulp, the major pollutants are bits and pieces of plant waste, such as bark and finely ground wood. More commonly, pulp-mill operation goes beyond the ground-wood method and proceeds through a variety of chemical treatments. In the soda process, the usual one for woods such as poplar, soda ash and lime are blended with wood chips and the mix is then boiled under pressure. Coniferous woods are more commonly treated by either the kraft or the sulphite method, both of which involve digestion of the wood in acidic calcium sulphate. After the digestion process, the waste "liquor" is usually not recoverable and is discharged. In the kraft process, digestion is with a highly alkaline sodium sulphide solution. The black liquor waste, which is separated from the fibres, can then be treated to recover a large proportion of the pulping chemicals, thus producing much less of a pollution problem than the sulphite process.

The toxic materials in pulp effluents are of a great many kinds and, indeed, all the individual components are not known. In kraft-mill discharge some of the toxic compounds are: hydrogen sulphide, methyl mercaptans (these two make most of the smells), sodium thiosulphate, various soaps and fatty acids, and resin acids. When fish are exposed to pulp-mill effluent, they generally show signs of respiratory difficulty and this is probably a result of the precipitation of proteins in the mucous on the gills. This is a similar response to that which occurs with heavy metals, and it may be one of the most common modes of action of aquatic pollutants. Some of the toxic components of pulp-mill effluent exert their action internally. For instance, methyl mercaptans appear to work directly on the central nervous system, causing death by paralysis of the muscles used for moving the gills.

Pulp-mill wastes also have a number of "sub-lethal" effects—
the fish are not killed, but are debilitated by chronic exposure.
For example, fish that have been exposed to pulp-mill effluents
do not have the same swimming endurance, they have reduced
breathing rates, heart rates, and growth rates. There are changes
in red-blood cell counts and in many other indicators of general
well-being. There are also shifts in behaviour: species of fish that
normally school may fail to do so in the presence of pulp-mill
effluents. They may exhibit a reluctance to eat and a variety of
other symptoms which may be indicative of abnormal behaviour
and physiological distress.

The extreme acidity or alkalinity of pulp-mill wastes may be
responsible for much of their toxicity. The pH of effluents may
range from 7.5 to 11.5 for alkaline processes, and from 4.5 to
6 for acidic processes. With such extremes of alkalinity or acidity, it is scarcely surprising that there could be such dramatic
effects on fish.

Most species of fish are sensitive to pulp-mill effluents and will
avoid areas where concentrations are high: chinook salmon
shun even low concentrations. By contrast, there have been
reports that fish are attracted to pulp-mill effluent, perhaps
because of higher temperatures.

Pulp mill effluents also affect fish indirectly by their influence
upon a variety of stream organisms. Much of the organic waste
is made up of the relatively insoluble lignin, and for ground-
wood process mills there may be a large production of bark and
finely powdered pulp. These materials smother and eliminate all
bottom organisms. In a study on the Spanish River in Ontario,
there was a scarcity of the usual stream organisms (midge larvae, aquatic snails, mayflies) for a distance of 20 miles downstream from the mill outfall. Most of this effect resulted from
the accumulation of fibre sediment that led to a reduction in
oxygen concentration on the bottom. Only species, such as
tubificids, that were tolerant of low oxygen concentrations could
survive.

Pulp-mill effluents may also be harmful to vegetation.
Aquatic plants may become covered with a coating of pulp fibre
to such an extent that the gas exchange in photosynthesis and
respiration is impaired. The discharge may also discolour the
water and contribute to its turbidity so that light penetration

is reduced. There are also some interesting physical effects. Pulp-mill outflows increase the specific gravity of water and the effluent therefore tends to sink. Some of the volatile components form a surface film that may prevent the development of waves or ripples. Some species of aquatic plants benefit from these changes and flourish under polluted conditions. Other species are, of course, crowded out. The result is that downstream from pulp mills there is usually a different association of aquatic plants.

Finally, pulp-mill effluents usually reduce recreational pursuits in the near vicinity. The combination of odour, discolouration of the water, and foamy residues is not conducive to a pleasant day of fishing or boating.

There are many ways in which pulp-mill pollution can be reduced. In more modern mills emphasis is given to processes in which the chemicals can be recycled. The techniques of making pulp are gradually being improved and there is less loss of fibre. Settling basins and other forms of treatment can move the pollution out of a river and, like a sewage plant, concentrate the biological activity where it does less harm to natural environments. Where the discharge from a mill is relatively small and the flow of a river is relatively large, the pollution effects are minimal. For example, where pulp mills are located on salt water there can be very rapid and effective diffusion of their discharge by strong tidal action: in these circumstances, there are only local problems, fish and other animals being affected just in the immediate vicinity of the outflow.

There has been a long history of pulp-mill pollution in Canada and it is still a serious problem in many local areas. Particularly in eastern Canada, each year more and more of the older mills are being phased out and are being replaced by modern facilities that are required to meet much more stringent pollution-control regulations. Nevertheless, pulp-mill pollution will remain a problem in Canada for many years, and there is much yet to be done to develop the technologies that might make this a truly pollution-free industry.

OTHER INDUSTRIAL POLLUTANTS

A wide variety of industries produce effluents that are complex mixtures of an assortment of chemical wastes. In some instances the effect of this cocktail may be severe only immediately near the point of discharge, and may be the simple consequence of a change to the acidity or alkalinity, a sharp increase in salt concentration, or a combination of these effects with a reduction in oxygen. It is simply "gross pollution" (see p. 45) on a small scale. Because these effluents are so commonly mixtures of wastes, it is seldom possible to pinpoint what part was toxic. As a consequence, the literature on pollution from many manufacturing industries is particularly complex.

The situation is made even more complicated because the chemistry of the receiving waters is also variable. An effluent may be effectively neutralized in some kinds of streams (typically, hard-water streams), but may be lethal in others. The temperature and oxygen concentration that are prevailing at the time also strongly influence the effects of pollutants. Finally, the actions of most toxicants depend on the nature of the receiving waters and also on the species of organisms present. In general, fish such as trout and salmon are far more sensitive to toxic materials than are goldfish or carp. Some species of aquatic insects are far more susceptible than others to effects of insecticides.

As a consequence, it is fairly common to hear apparently inconsistent statements about how toxic some material may be. The reports only begin to make sense when the circumstances are clearly defined—the species of organism, the chemical nature of the water, the duration of the period of exposure, the temperature, even perhaps the season of the year (because spawning or young fish may be differently sensitive), and so on. Despite the foregoing, there are some substances that can be singled out for special mention because they are toxic over a wide range of situations to a large number of aquatic animals.

Phenol, for example, is produced as a waste in a great variety of industrial operations. It is used in several kinds of glues and disinfectants, is a waste from coking operations, and has many applications. In concentrations of from 1 to 20 ppm, phenol is directly toxic to fish and other aquatic organisms, and in the ordinary procedures of housekeeping it should be relatively easy to keep effluents down so as to avoid these intensities. Unfortunately, though, phenol imparts a distinctive carbolic-soap taste to fish flesh even at concentrations as low as .01 ppm. It is alleged that on the river Rhine the few Atlantic salmon that were caught as the species was slowly eliminated had this unpleasant flavour. Phenol breaks down fairly rapidly in fresh water but, even so, the odour is so highly distinctive that it can often be smelled in the fine mist of a waterfall many miles below an industrial discharge, and it persists as a faint flavour in the flesh of fish for even greater distances downstream.

Ammonia is another common industrial waste that is directly toxic to fish. Unlike many other toxicants, its effects are most severe in alkaline waters with high carbon dioxide content. Ammonia may also be a common ingredient of farm and stockyard wastes. It has sub-lethal effects on fish at 0.1 ppm, decreasing the oxygen-carrying ability of the haemoglobin in the blood. Concentrations of over 2 ppm are lethal for some species of fish.

Cyanide is another common pollutant that is a well-known poison and is used in a variety of chemical industries, particularly when metals are being processed. Cyanide in its ionic form is lethal to trout at concentrations of only 0.1 ppm. It is equally toxic at low concentrations to a great many kinds of organisms. It is usual for cyanide to be released in association with various compounds with which it forms complexes that have different toxicities to fish. For example, combinations of zinc and cyanide, or cadmium and cyanide, are extremely toxic. In acid waters, the nickel-cyanide mixture breaks down readily and free cyanide is released, but in alkaline waters it is much less toxic than other cyanide complexes. Accordingly, adding nickel salts to cyanide wastes can be a way of reducing toxicity of cyanide in a crisis situation.

Cyanide is a fairly common pollutant in Canada because of our large production of metals. It is a regular constituent of smelting-plant wastes, and also shows up in the vicinity of large cities where there are metal electroplating operations and other industrial uses.

Arsenic is also commonly produced as a waste product in smelting operations, and in a variety of manufacturing processes such as tanning and dyeing. It is, of course, dangerous to humans, and is an accumulative poison. For this reason, discharges of arsenic should be kept at very low levels if water is to be used for human consumption. In general, fish and other organisms are also susceptible to arsenic poisoning, though they may be able to withstand concentrations of as much as 1 to 2 ppm. Sodium arsenite is used as a herbicide for aquatic plants and can be applied up to concentrations of 10 ppm without killing fish. Nevertheless, arsenic is a dangerous contaminant because it may accumulate; thus we should probably consider all levels above 1 ppm as potentially dangerous.

Many other compounds have particular toxic actions and among those worthy of passing mention are fluorides, elemental phosphorous, nitrates, benzene, and cresols. In recent decades a multitude of organic chemicals have been added to the list of pollutants, and for many of them it is not known what their possible effects are. Suffice it to say that either the level of research should be greatly increased, or their discharge into natural lakes and streams should be prohibited until they are proven harmless.

chapter six

EUTROPHICATION

All lakes and streams undergo a natural "ageing" process. As lakes gradually fill, they become more productive, or eutrophic; further downstream, watercourses are generally slower moving, richer in nutrients, warmer, and hence more productive or eutrophic. When man intervenes to accelerate these processes, we say that lakes and streams suffer from "eutrophication." It would perhaps be better to say thay they were "culturally influenced," because the way in which they are affected is not identical to the natural process. Man's influence usually takes the form of excessively increasing the inflow of particular elements (such as phosphorous), and augmenting the flow greatly over a short span of years. As a consequence, a dislocation of natural processes results in conditions that do not occur in the usual ageing or eutrophication process, and the character of the dislocation depends on what is being added and in what quantities.

The commonest cause of eutrophication is the addition of

phosphates to lake waters, because the rate of production of aquatic vegetation is usually limited by the small amounts of phosphorous in the water. As anyone knows who has a garden, the same is true on land, and if you want to make grass grow, add phosphate fertilizer. But phosphorous itself will only stimulate growth when other nutrient elements are available. Typically, nitrogen is also in relatively short supply, at least for those organisms that cannot "fix" nitrogen. Carbon may also limit the rate of biological processes in waters of very low alkalinity. Many other elements are necessary in small or trace amounts, and their proportions can influence which plants grow most effectively. The addition to a lake of phosphates alone may thus have an effect different from that following the introduction of a more "balanced fertilizer" such as human excrement, and neither gives the same reaction as does the combination of both phosphates and excrement in the form of a discharge of sewage. Lake and stream waters also vary greatly in the amounts and proportions of different mineral nutrients they contain. The same effluent added to two lakes may have different effects.

In these circumstances, it is difficult to pinpoint what particularly is responsible for the observed eutrophication of a lake. Usually the cause is multiple—the addition of phosphates, nitrates, sewage, and other materials—and the effect is a gross and fairly rapid deterioration of natural conditions.

Lake Erie is the prime example of eutrophication in Canada and it has often been called a "dying lake." This condition has come about from the addition to the lake of phosphates from detergents, phosphates and nitrates from agricultural fertilizers, quantities of silt from eroded watersheds, human sewage, and a mix of industrial effluents of various kinds. Each day, Lake Erie receives 20,000 pounds of phosphates and each pound of phosphate can grow 700 pounds of algae. The many additives have overtaxed the natural assimilative capacity of the lake, fostered intense long-lasting blooms of algae, and created the unusual phenomenon of an oxygen-deficient hypolimnion in a large, deep, temperate lake. In an area of 2,500 square miles, there now is virtually no dissolved oxygen in the water within ten feet of the bottom.

Lake Erie is the most seriously affected of the Great Lakes, but Lake Ontario is also severely influenced. The total dissolved solids have increased from 140 to 190 ppm since the turn of the century, and most of this increment has been since World War II (figure 19). Lake Michigan is the next most influenced, with an increase from 130 to 150 ppm. Lake Huron and Lake Superior are least affected to date. Recently, all of the lakes have become the sites, or proposed sites, for nuclear reactor power plants that would cause a temperature increase in the lake waters and further accelerate eutrophication.

The totality of environmental changes, plus the effect of the fishing industry, has seen the virtual elimination of the historical fisheries on Lake Erie (figure 20). Whitefish catches have declined from 2,000,000 to 13,000 pounds (1962), cisco catches from several million to 7,000, sauger from 1,000,000 to 4,000, walleye from 15,000,000 to 1,000,000, blue pike from 15,000,-000 to 1,000. The total production has remained more or less constant, with carp, yellow perch, drum, and smelt replacing the finer species in the catch. But, with further deterioration, these species, typical of eutrophic lakes, would also be eliminated. Fortunately, the whole process is largely reversible. If phosphate introductions are reduced, and if there is curtailment of other effluents, the lake will return to a state much as it was previously, except that the bottom sediments will have recorded the years in which the natural processes were disturbed. With complete control of discharges, Lake Erie could be back to natural conditions in five to six years. Some progress is being made in this direction now as Canada and the United States conduct a joint program aimed at rehabilitating Lake Erie and the other Great Lakes.

On a smaller and less dramatic scale, there are hundreds of examples of eutrophication of Canadian lakes. Modern agricultural practices rely heavily on phosphate and nitrate fertilizers; though most of the additions are bound to the soil particles there is some leaching, especially when the fertilizer is applied too heavily. Forestry practices also contribute to eutrophication. In their natural state, forests trap most of the nutrient materials before they can escape into streams and lakes. Vegetation also acts as a natural sponge, soaking up a large proportion of the rainfall. When trees are cut, there may follow a period of excessive decomposition of waste (needles, bark, twigs, sawdust) coupled with high run-off of water. There may also be soil erosion, especially on steep slopes. As a consequence, the streams and lakes are suddenly charged with much greater than usual quantities of organic and mineral elements and the result is a temporary, small-scale eutrophication.

There are also many local effects that arise from the seepage from septic tanks of cottages that line the shores of our beautiful and accessible lakes and streams, or from such thoughtless acts as creating garbage dumps on lake shores or stream banks. In special circumstances, these local cases of eutrophication can be dramatic. For example, one of the small lakes near Resolute Bay has clearly shown the impact of a camp of geologists who inadvertently enriched the otherwise nutrient-poor inflow into an arctic lake. A small lake near Princeton (B.C.) has developed a flocculent bloom of algae and bacteria because of the inflow of wastes from summer visitors. The mobility of modern man redistributes the world's nutrients!

Not all of these effects should be considered as necessarily harmful. There is much in nature that can be improved upon. For example, at Great Central Lake on Vancouver Island, the Fisheries Research Board of Canada is engaged in research designed to increase the production of sockeye salmon. Basically, the approach is to fertilize the lake with phosphates and nitrates, and thereby to augment production of phytoplankton, and thus production of zooplankton, and, finally, to therefore increase growth rate and decrease mortality of young sockeye salmon that stay in the lake for a year before they go to sea. As might be expected, the British Columbia Pollution Control Board was at first against the proposal because it meant adding phosphates!

Some time in the future we will probably be controlling the rates at which various elements are added (and subtracted) from the natural waters of lakes and streams to obtain maximum advantage for man. Meanwhile, we are commonly involved in trying to prevent uncontrolled and unpremeditated discharges that have generated eutrophication effects that were not anticipated.

chapter seven

WASTES FROM POWER PRODUCTION— HEAT AND NUCLEAR MATERIALS

WASTE HEAT

A large number of industrial processes require great quantities of water for cooling purposes. The effluent from these kinds of operations is hot water—water in which the pollutant is waste heat. Obviously, when water temperatures reach extremely high levels all living organisms are unable to survive and, for this reason, problems of disposing of hot water have become increasingly severe with growing development of industry. Power plants, particularly those that are based on nuclear reactors (which require large quantities of cooling water), are a more recent source of concern, and a great deal of thought has been given to ways in which waste heat can best be dissipated from large power installations. The problem has particular urgency for the next few decades as power demands escalate. For example, it has been estimated that, by 1985, more than 75 per cent of the surface and sub-surface water in the United States will flow through cooling condensers of one kind or another. An obvious technique is to allow steam to flow off into the atmosphere, but this can create serious local weather conditions downwind from the cooling ponds or cooling towers. By contrast, it appears likely that waste heat can be discharged into rivers and lakes in limited amounts, and may even have beneficial rather than harmful effects.

It has long been known that small increases in temperature have beneficial effects on the organisms that inhabit temperate waters. Provided the warming is not sufficient to cause the death of the various organisms, the usual effect is to speed up all biological processes, increasing production and raising the rate of turnover. Thus, in a warm summer, a lake or stream will tend to be more productive than in a cool one, other things being equal. The key, then, to dealing with waste heat is to disperse it adequately so that it nowhere reaches such high levels as to be a dangerous "pollutant."

In the immediate vicinity of an outfall from a large power plant or industry the water temperature in a plume from the discharge point may be as much as 100° more than is normal. But, in a relatively short distance there is usually sufficient mixing that the whole of a river may be warmed by only a few degrees. The consequence is that immediately downstream from an industrial outfall of hot water, a small number and a small diversity of organisms can tolerate the extremely high temperatures, but beyond this lethal zone there is increased variety and abundance of all kinds of aquatic organisms. The skilful engineering of hot water discharge can thus be made to have beneficial effects.

Where the discharge of heat is large in relation to the size of the river, and where there are several outflows into the same stream, the total effect can of course be disastrous. Not only are conditions intolerable for the organisms some distance downstream, but any migrants in the stream are blocked or killed in passage through the high-temperature zone. If, at some times of the year, the flow is extremely low, then the temperature effects become exaggerated and all of the high production that has been created by the increase in temperature may be destroyed in a short period. Obviously, the successful dissipation of waste heat to produce beneficial results involves development of a year-round plan of operation for disposal with a large built-in safety factor. Rates of discharge must be carefully related to natural flow and should be patterned so as to have the best possible effects.

The addition of heat to a lake can also have beneficial results, although the patterns of lake circulation can be immensely complicated by introducing hot water at one spot. If the hot water is added at the surface, it speeds up the development of a stable epilimnion; if at the bottom, it creates vertical convection currents and locally distributes the heat over all depths. Where the hot water warms one side of a lake, currents are induced (probably alongshore) that eventually distribute the heat more evenly. Regardless of the way in which the hot water is added, there are potentially profound effects on physical structure and circulation in the whole lake. The depth from which cooling water is drawn can also influence the physical structure, regardless of whether the hot water is returned to the lake. If water is taken from the hypolimnion, the lake develops a much thicker epilimnion and this can have harmful effects on cold-water species, especially in the summer. On the other hand, if the water is drawn from the epilimnion, a lake may be cooled slightly at the surface with, in some situations, beneficial effects, and in others slightly harmful results. Certainly, any scheme for using lakes as sources of cold water or receivers of hot water should be preceded by a thorough study of physical limnology to enable assessment of its effects on circulation. The species of organisms are usually different in the areas that are warmed and, in general, productivity may be higher. As with streams, there seem to be possibilities for some imaginative engineering that would achieve rapid distribution of waste heat throughout the whole of a large lake to produce uniform and substantial increases in productivity.

It has been widely suggested that waste heat plus sewage are the major ingredients for successful greenhouse operation. Ideally, one might build a power plant immediately adjacent to a sewage treatment plant and use the wastes of both in greenhouse gardening. It has been further proposed that there might be possibilities for developing a culture of many species of fish in these kinds of "farms." Most attempts along this line to date have been on a relatively small scale, but they have all indicated that there are substantial potentials to be developed.

Unlike most waste commodities that go into water, heat seems to be something that we not only can do something about, but can take advantage of. Atomic Energy of Canada is currently developing research in this field and is trying to evolve engineering skills that would be useful all over the world.

NUCLEAR WASTE

Almost since the time that radioactivity was discovered, it has been known that radioactive compounds are potentially harmful to living organisms. With the development of the atomic bomb and the perfection of the technologies of nuclear power, it has become everyday knowledge that these compounds are potentially the world's most dangerous pollutants.

Like all other pollutants, radioactive compounds appear to have minimal effects at low doses. We are all exposed to natural radiation levels in the order of 0.1 rems per year.* This comes from cosmic rays and from naturally occurring radioactive compounds in rocks and in the soil. It is presently generally accepted that the level of radiation can be as high as 0.5 rem/year without endangering health, although there may well be some subtle mutagenic and carcinogenic effects that are difficult to measure. People who work in nuclear plants, or who work with radioactive compounds, are generally not allowed to be exposed to more than 5.0 rems/year. All these levels are relatively safe standards, according to the best modern knowledge.

It would thus seem to be a relatively straightforward matter to engineer nuclear power plants so that the levels of radioactivity in the effluent were always so low that the general public would not be exposed to dangerous doses. To take an extreme case, if the effluent from a nuclear power plant could be uniformly spread over the whole world, the net effect would be

*"Rem" is the name of the unit used to describe the biological effects of radioactivity regardless of the source or kind of radiation. To say that a person has received a dose of a certain number of rems is a way of describing how much effect the radiation has had.

virtually unmeasurable. The object, then, is to so dilute the effluent that the level of radioactivity is quickly reduced to low levels. By delaying the release of cooling water from a nuclear plant, most of the radioactivity that comes from compounds with a short half-life can be almost completely dissipated. The more extreme radioactive compounds can be taken out of the cooling water and put into permanent storage in specially constructed dumping grounds. The discharge from a nuclear plant can thus be brought well below the hazard level where there need be public concern.

At this point the plot becomes a bit more complicated, because many organisms concentrate radioactive materials; for example, phytoplankton intensify radionuclides 200,000 times, insect larvae and fishes 100,000 times. The yolk of duck eggs may contain radionuclides concentrated to 1.5 million times the level found in the water in which they live. This assembling of radionuclides can pose serious health hazards to anyone who takes his drinking water from a contaminated area, or who eats a fair amount of plant or animal matter that lived there. The problem is further complicated in that the size of the area affected relates in part to physical circulation and in part to the level of productivity. If radioactive compounds are discharged into a region of low productivity, then the rate at which they are taken up is low and they will tend to be dispersed over a large area. If productivity is high, then the nuclides are taken up very rapidly and can become concentrated in a small zone. Thus, to take an extreme example, if one were to discharge sewage immediately adjacent to the effluent from a nuclear plant, it would be expected that the sewage plus the waste heat would greatly enhance productivity and create a small area in which there was a very high concentration of radionuclides in the plants and animals.

People closest to nuclear plants will obviously get the biggest exposure and, if it is an area of high productivity, their dose could be elevated if they drink the water and eat the fish. Should the power plant be designed so that no *individual* is affected, or would it be permissible to provide protection for the average individual? If the average, do we include all those who benefit from the power?

There are many social implications to this situation. Is it

better to seriously influence 1 per cent of the people than to influence everybody by 1 per cent? To my mind, the answer is that it is better to affect no one at all, but this apparently poses enormous economic and engineering problems that could in effect virtually deprive us of the benefits of nuclear power. And what are the alternatives to nuclear power? Do they pose problems of pollution that are of more serious concern to public health and aquatic organisms? When two nuclear plants are close together, how will the limits for effluent discharge be defined? There will be an extensive public debate on these subjects in the next decade in Canada and elsewhere in the world as nuclear power is increasingly seen as the only feasible way of maintaining a high energy production. Nuclear power has already been developed at several sites in Canada, and there is every prospect that there will be demand for the construction of many more such stations (figure 21).

chapter eight

LAND USE, WATER USE, AND POLLUTION

The foregoing chapters have indicated that virtually any substantial change in the characteristics of freshwater lakes and streams may have harmful effects on aquatic organisms and aquatic productivity. When a pollutant is discharged from industrial plants, we can easily see that it has an immediate effect where it is poured into a stream. Where a sewage outfall directs its effluent into a lake or stream, the polluting impact is equally obvious. It is these sorts of visible damage that we usually associate with the word pollution. But they are by no means the only kinds, nor are they necessarily the most severe. Lakes and streams are very much a product of the land areas they drain. Virtually any pattern of use of the land is reflected in the characteristics of the streams and lakes.

For example, the extensive logging of many parts of Canada has had substantial effects on lake and stream biology. When trees are cut down over large areas, the ability of the land to absorb moisture is reduced. As a result, there is much less control over the natural run-off. A natural stream in a heavily forested area will have neither a large spring flood nor a very low flow in the summer. The ability of vegetation to hold and absorb water means that any large fluctuations in rainfall are smoothed out. Once the tree cover is gone, a heavy rainfall is quickly reflected in a rise in stream level. Severe freshets in the wet months or in the spring may change the stream channel and create an unstable environment for fish and other organisms. In extreme situations the gravel in a stream may be ripped out by a heavy flood and the eggs of fish such as salmon completely destroyed. Additionally, many of these watercourses may go virtually dry in the summer and their resident populations of trout and young salmon are eliminated. In many west-coast streams the decline in salmon populations can thus be traced to the effects of logging on the pattern of stream flow.

The removal of forest cover may also result in soil erosion and changes in the chemical characteristics of the run-off. Many recent studies have indicated that the chemistry of stream and lake waters is substantially changed after logging in the watershed, particularly if the slash is burned. In general, the quantities of nutrients are much greater and logging therefore usually contributes to eutrophication. Another effect is that stream temperatures generally rise because of the absence of shade from trees on the banks. The warming trend can be quite remarkable and in only a few miles a cold mountain stream can quickly gain sufficient heat to make it no longer a suitable environment for salmon and trout.

The construction of logging roads and the removal of tree cover may also greatly accelerate processes of stream erosion, with the consequence that these watercourses become heavily silted. The effects are usually to greatly reduce stream production and to render the bottom unsuitable for the spawning and rearing of fish (figure 22).

Thus, though it is far more diffuse, logging can have a devastating effect on stream biology—a far greater impact, perhaps, than a sewage outfall or a small industrial pollution. There are examples of the results of logging in all of the forested areas of Canada. The streams of New Brunswick, the drainages on the north shore of the Great Lakes, the many complex drainages in British Columbia, almost without exception, have been substantially modified by the impacts of logging.

Although it is much less common now than it used to be, the practice of driving logs down rivers also has many harmful effects on streams. Logs bounce on the bottom and scour stream gravels, and jams can create major obstacles to fish migration (figure 23). The damage to logs that results from their bouncing on the bottom and the blockages caused by log jams usually encourage loggers to make changes in the stream so that it will serve their purposes better. For example, they may build dams at the outlets of lakes to store water that can be used to flush the logs down the stream. Any of the sharp bends and twists in a river can be eliminated in a few hours with a bulldozer, and with a much straighter channel the risks of log jams are diminished. A stream can thus be reduced to a sluiceway for logs.

Logging practices in Canada in the past have largely been characterized by a singlemindedness that is perhaps appropriate to the opening up of new country. Once the virgin crop has been taken and logging must depend on what production can be sustained, the whole character of the operation must be changed if it is to persist. Trees must be farmed rather than hunted; increasing attention must be given to proper land husbandry. In essence, good forestry practices will automatically mean good water management, for where trees are properly farmed watersheds will be stable. Moreover, it is increasingly apparent that Canadians wish to use their forests for other purposes than just as sources of wood. The growing concentration of population in large cities is associated with a mounting desire for recreation in peaceful natural settings. Forested land must therefore be managed more and more with aesthetic considerations in mind. By degrees this is happening in many parts of Canada, and it is much to be encouraged.

No discussion of forest land use would be complete without reference to our wasteful ways with wood and wood products in Canada. Partly perhaps because we have so much forest land and partly perhaps because we are unthinking, we conduct our forestry and our everyday lives in a state of mind that is almost euphoric. Large quantities of wood products are left to rot in our forests because it is ostensibly uneconomic to harvest them. Quantities of wood fibre and waste are discharged from every pulp mill. Paper bags are seldom used more than twice in

Canada: once when they are brought home from the store and once (perhaps) as they go to the garbage pail. Most of us are unduly influenced by packaging, and we accordingly pay a higher price for our groceries. These are things that we all realize. What is less commonly appreciated is that in these many ways we are rapidly bringing the day closer when our forested land will be less able to maintain our prodigal habits. Our sustainable forest production in Canada is now approaching its upper limit, and in some parts of the country it has been surpassed. Is waste a wise policy now?

Logging is only one example of the way in which land use can influence lakes and streams. There are many other human activities that can have equally profound effects. For example, the construction of roads, whether superhighways, backcountry trails, or logging access routes, can have quite serious impacts on streams and their fishes. Where gravel is taken from streams for construction purposes, the result on local drainages can be very marked. In many parts of Canada the obvious place to look for gravel is in a stream bed, particularly where it enters a lake or the sea and there is a fine fan of clean, washed, and, to some degree, sorted sediments. The effects on stream organisms are of course disastrous, and the worst scars of road construction are frequently several miles from the site of a road. Where roads pass across watercourses, the way in which the culvert is placed may create an obstruction to the migration of fishes. The diversion of streams into new channels can cause soil erosion and, of course, dries up portions of the original bed. In a multitude of small ways, careless road building can have serious effects on streams and their fish inhabitants. There will be many problems of this kind when the Mackenzie pipe line and highway are constructed.

Agricultural practices are no less a factor in giving lakes and streams a character that differs markedly from their natural state. The addition of chemical fertilizers results in some leaching to lakes and streams, and contributes to their eutrophication. The use of herbicides, insecticides, and fungicides is conducive to a steady pollution which is difficult to control because it comes from no single source. Basically, it is the way in which agriculture is conducted that influences lake and stream biology. The increasing industrialization of agriculture implies more widespread and systematic use of chemical fertilizers and pesticides. Concentration of livestock into small areas is a more efficient way of producing meat, but it involves an intensification of waste that inevitably leads to pollution. As with forestry, agriculture is competitive in world markets, and to become more efficient it becomes more singleminded and industrialized. Aside from all the other ecological consequences of this trend (greater likelihood of pest outbreaks, soil deterioration, and so on), modern agriculture is substantially influencing the character of the lakes and streams that drain the land. We are farming at the expense of the countryside.

In areas where precipitation amounts are small, irrigation practices may also be a factor in influencing stream biology. The diversion of streams into irrigation canals almost invariably means that large numbers of fish eventually die when they become stranded in irrigated fields. Excess irrigation water ultimately drains back into streams; it usually is substantially contaminated with chemical fertilizers and pesticides. Moreover, there is the effect of the removal of water from a stream, ordinarily at the times of year when stream flows are already low. In these circumstances, irrigation can virtually eliminate aquatic environments. Three-quarters of the streams tributary to Okanagan Lake in British Columbia are no longer spawning and rearing areas for fish for just these reasons. To augment summer flows for irrigation, it is not uncommon to use a small lake as a reservoir with a dam and control valve at the outlet. Most of these situations have some good features, but mostly bad ones. Although they at first tend to stabilize both lake and stream levels, and in this way enhance aquatic productivity, they generally encourage much more ambitious irrigation schemes so that ultimately the streams are as dry as they would have been without the reservoir, and the latter has a greater fluctuation in level than the lake it replaced. Irrigation schemes should be conceived so as to maximize the benefits both for productivity of crops on land and for the maintenance of aquatic ecosystems.

Excessive irrigation may also contribute to making soils and streams much more saline than they are naturally. If land is flooded, the natural salts in the soil became dissolved. When the water subsequently evaporates, the salts are precipitated and an alkaline crust is formed on the soil surface. The problem may be more severe if the irrigation water is from an alkaline source.

The foregoing are only some broad examples of the ways in which land use can affect lakes and streams. For virtually every major human activity on land, there are consequences on lakes and streams. Much of our success as resource managers will hinge on ensuring that what we do on the land will not by inadvertence destroy what exists in our waters.

Major dams for power development for industry, or even for large-scale domestic purposes in big cities, also have a serious impact on the biology of lakes and streams. A large dam usually converts a portion of a stream into a reservoir that superficially looks like a lake but which, biologically, is quite different. For example, most reservoirs have a shape which is unnatural. There may be arms that reflect the outline of the natural drainage basin that is being flooded but, in general, reservoirs are long and thin, and odd in that their outlet is virtually a sheer vertical face. The physical circulation in reservoirs is usually peculiar. If water is drawn from the bottom of the reservoir, the zone of warm water on the surface becomes extremely deep and normal processes of production in the upper zone and decomposition in the lower zone can be radically altered. If water is drawn from the surface, the upper zone may be extremely thin and not so productive as it would be otherwise. To maximize aquatic production in reservoirs it is necessary to manipulate discharges so as to achieve the optimum thickness of epilimnion.

A much more drastic physical characteristic of reservoirs is their wide fluctuation in level. The whole idea of building a dam on a river is to store the water at periods of heavy run-off and to release it at times of low flow, thus smoothing out the usual fluctuations in river level. This, of course, can only be achieved if the level of the stored water goes up and down. In most reservoirs there is an extreme change in water level that has the effect of eliminating production all around the shoreline. Because these areas are generally the most fruitful in a lake, it follows that reservoirs generally have low levels of productivity. Changes in water level can also be extremely harmful to those species of fish which commonly spawn near shore. Consequently, reservoirs generally have quite unusual fish populations, and most of the species which can persist in them are not particularly useful for sport. Finally, fluctuations in level make reservoirs unsuitable for many recreational purposes. Most of the exposed shoreline is covered with fine silt or organic material, which is no substitute for natural beach. (Some dam builders give this fine silt the euphemistic description "sandy coloured.") Where possible, reservoirs should be maintained at their full height for the summer period to ensure optimum levels of productivity and the best characteristics that can be salvaged for recreation.

By far the most dramatic effect of constructing a dam in a river is its impact on migratory fishes. If a dam is less than 100 feet high and the river flow is relatively small, it is usually possible to build fishways or ladders that will enable the fish to safely bypass the dam. It is necessary, of course, that such devices should be designed professionally because it is very easy to make mistakes and the fish will not use what has been provided for them. With this proviso, in general, low dams on small rivers pose rather few problems for migratory fishes.

When a dam exceeds 100 feet in height, or the water flow becomes substantial, such as in a major river like the Fraser in British Columbia, a multitude of new problems are encountered. In a wide river the fish may expend a great deal of time finding the entrance to a fishway. This would perhaps be of little consequence if it were not for the fact that migrating fish, such as salmon, do not eat and they have limited reserves of energy to make the journey to their home stream. If they lose ten days looking for the entrance to a fishway, when they reach the spawning grounds they may die without depositing their eggs. This has been observed in many situations where there have been natural obstructions such as rock slides. Additionally, on large rivers the populations of migratory fish may be very large; thus even if they can find the fishway quickly they may be delayed because only a certain number can pass at a time. Salmon, for example, will not pack themselves into a fishway shoulder to shoulder; if a large number is to go through in a hurry, then a relatively large fishway must be provided. The same is true of an elevator system, a fish lock, or any other device in which space is restricted. The only answer to this problem is money, and the provision of suitable fishways may be prohibitively expensive.

The problem of migration of adult fish upstream over dams does not end when they leave the fishway. In the waters of the reservoir the fish may become confused and be washed back over the spillway to do the whole climb once again. The waters of the reservoir generally do not provide a good orientation for direction and the fish may spend a considerable period of time milling about before proceeding upstream. Any delays are thus aggravated. Even if a perfect fishway is designed, there is reason to doubt that the problems of adult migration will be solved. On this score alone high dams on large rivers are incompatible with migratory fish like salmon.

Dams, of course, also have their effects on young fish going downstream. The passage to the sea for the young fish may be delayed in the reservoir just as the upstream passage of the adult fish was retarded. Many of the young fish may never get out of the reservoir because the physical circulation may be such that they do not receive the necessary environmental cues to lead them to the outlet at the dam. Once they reach the dam, there are hazards whether they go out on the surface or into an orifice at some depth (figure 24). In the route over the spillway young fish may be killed by abrasion on cement surfaces, or by the impact when the spill strikes the tailwater at the foot of the dam. In coming through a deep orifice, there are the risks of pressure changes associated with the turbine blades of power generators and with the movement from deep water to the surface. Additionally, water immediately below the dam may become supercharged with dissolved air and, in these circumstances, fish may suffer from what is called "nitrogen bubble disease" which is very similar to "the bends" that affect human divers when they come to the surface too rapidly. Many of the losses to salmon in the Columbia River in the United States have occurred for this reason. Considerable effort has been expended on devising ways of diverting young salmon so that they leave a reservoir by a safe bypass—sort of a fishway going down. The fish will certainly respond to a variety of things and can be guided by such devices as electric currents, bright lights, or streams of air bubbles. In addition, it is sometimes possible to screen the spillway so that there is virtually no way out except for the safe bypass. But, as with fishways and fish ladders, on very large rivers these sorts of solutions become enormously expensive, if not mechanically impossible. For any dam that is several hundred feet high, there can be very little optimism about the survival of the young fish coming downstream. When these considerations are added to the problems of adult migration, there can be little doubt that building large dams on major rivers clearly means the sacrifice of migratory fish populations.

Construction of large dams may also have far reaching effects on aquatic ecology many miles downstream. Where a lake takes the place of a stream, there is much greater warming of the water, and there also is heat generated in the process of developing electrical power. A series of dams on a major river can thus lead to an appreciable increase in water temperature. Reservoirs also act as a settling basin for fine silt in rivers and this again means a change in the natural characteristics downstream. It has been argued fairly persuasively that some large dams have influenced the ecology of estuarial regions of rivers. Undoubtedly, these effects occur but, to date, there have been so few detailed studies that it is difficult to predict whether such consequences will be beneficial or harmful to estuarial organisms.

A rather special example of downstream effects occurred with the construction of the W. A. C. Bennett dam on the Peace River in British Columbia. The natural regime of stream flow of course involved a large spring flood. This annual inundation was responsible for flooding the marshes at the west end of Lake Athabasca and each year created an excellent resting and nesting area for geese and ducks, an outstanding habitat for muskrat, and a splendid environment for goldeyes (which are a commercially valuable species of fish). With the curtailment of the spring flood, the whole area rapidly deteriorated as a wildlife and fish producer. The construction of a device which creates an ice jam and a flood every spring, or the building of a weir, may do much to restore the area to its original condition. Because corrective measures have not been in operation for a sufficient number of years to have had an adequate trial, it is not yet certain that any of the schemes so far envisaged will do the job. The effects of the W. A. C. Bennett dam also extend further downstream, and the level of Great Slave Lake will in the future be lower by about a foot.

There are many variations on the basic themes that have been mentioned above. For example, some dams built at lake outlets are primarily for storage, intended to assist in flood control, as well as to provide a regulation that is useful for power development further downstream. In these situations a dam may or may not obstruct migratory fish, but it invariably has a major impact on the lake basin which is flooded. The amount of water that can be stored in a basin enlarges dramatically with an increase in the height of a dam. For this reason it is usually tempting to construct a dam that raises the height of a lake well above its natural level. This, of course, means flooding of the shoreline and the creation of a lake whose margin, if left uncleared, is a jungle of flooded-out dead trees. Even if the nearshore vegetation is thoroughly and carefully removed, the new regime of fluctuation in level usually means the replacement of natural beaches by a quite muddy and unsightly foreshore. Some of these aesthetic considerations can be minimized if shifts in lake level are small during the summer months. In general, though, reservoirs created from natural lakes are not so good for most recreational purposes as the lakes they flooded out.

Reservoirs that have been built to date have largely been designed to store water that is in a watershed, and where there has been diversion to other watersheds it has usually been on only a local scale. Thus, the Aluminum Company of Canada dam on the Great Circle lakes (what was once Tweedsmuir Park in British Columbia) diverts a portion of the Nechako drainage from the Fraser to the Kemano, which enters the Pacific Ocean at Kitimat. This dam has changed the flow characteristics and the temperature of the Nechako River, and there have been effects on salmon, but as a diversion scheme it is a relatively local affair. By contrast, some of the proposals for water diversion in the future are continental in scale. For example, it has been proposed that much of the Yukon fresh water could be diverted to the water-poor regions of the southwestern United States, using the Rocky Mountain Trench (a fine U-shaped glacial valley) as a natural sluiceway. This would be part of a huge scheme called the "North American Water and Power Alliance" (NAWAPA), other features of which would be diversion of the Peace and Athabasca systems into the Great Lakes via a Canadian Great Lakes Waterway, and shunting much of the natural drainage usually going into James Bay instead to the south through Lake Huron. There are many variations of the

total scheme, involving such things as canals from Great Bear
Lake to Great Slave Lake, then on to Lake Athabasca, Reindeer
Lake, Lake Winnipeg, and from there splitting into two routes:
part of the water to go east to the Great Lakes and part to go
south to the Missouri and the Rio Grande (figure 25).

Although this total scheme may never be implemented, it is
without doubt feasible. Plans for the drainages into James Bay
came to a head in 1971 with announcement of the James Bay
Hydroelectric Power Development Project. This endeavour,
now in the early stages of construction, is a multi-billion dollar
series of ten dams that would influence the drainage of well over
100,000 square miles, about one-quarter of the area of Quebec
(figure 26). If the entire scheme is carried out, 6,000 square miles
of land would be flooded. Aside from the effects on the native
peoples, the total impact of a project of this size on freshwater
resources will undoubtedly be very large. It is astonishing that
there were virtually no ecological studies of the area before the
commitment was made to undertake the project. Some investigation is now under way, but it will almost certainly be too little
and too late. It seems probable, too, that whatever findings
emerge from these belated investigations will appear too late to
change the main plan of the project, and the "solution" of any
problems will be a series of patches, compensations, and hopes
that somehow the passage of time will either heal the scars or
get people used to them.

It must be realized that water-diversion schemes on these
mind-boggling scales are fraught with enormous biological and
social consequences. Primarily, they represent a major step in
technology that would transform the natural face of the continent. In essence, they would apportion the problems of some of
us to all of us. Many of the normal patterns of distribution and
production of freshwater organisms would be man-modified. It
might become increasingly necessary to seek recreation in less
natural surroundings. Large numbers of people would be dis-
placed from their traditional homes and ways of life. All of this,
it may be argued, is necessary if the continent's potentials are
to be achieved, but when these kinds of steps are taken there is
a commitment, at least for several hundred years, to an integra-
tion that may have wide repercussions. Major water-diversion
schemes of such a large scale could dwarf all other considera-

tions of freshwater management as principal determinants of Canada's future. It is obviously essential that we should make our decisions with great deliberation.

The foregoing are only examples of the ways in which the uses of land and water can influence aquatic environments. Although it is obvious, it is not generally realized that most lakes and streams are being affected in a multitude of ways that reflect the total impact of human activity. For example, the deterioration of Lake Erie was not due simply to the phosphates from detergents; it was caused by the addition of phosphates, from detergents and a number of other sources, combined with the intrusion of sewage, combined with loading with silt because of soil erosion in tributary drainages which was in turn a consequence of forestry practices. All this was amalgamated with industrial pollution, an intensive and poorly managed commercial fishery, and the gradual accumulation of the effects of a variety of pesticides.

In a similar way many of the major drainages of Canada are changing, not so much from the effect of any single human activity but because of the totality of a great many human activities. A single dam on a river may not pose too serious a problem for conservationists to solve, but the *existence* of one may make others more economic and the impact of a whole series of dams can be considerably greater than the sum of their effects would be if they were each individual dams. When a series of dams is combined with schemes for irrigation, when there is logging and agriculture in the watershed, it is inevitable that a river system will take on an entirely different character from that of its natural state. Moreover, any river in which many of the natural resources have been developed almost invariably supports a substantial human population, usually in the lower portions that are accessible for navigation from the sea. The mouth of the river then becomes a critical area. It receives water from a river system that has already been substantially modified, and adds to it a host of discharges of sewage and industrial waste. The whole mess is then subject to the blocking effects of tides on circulation and can create major ecological problems.

It has been said that rivers are integrators of the effects of human activity, and this is indeed true. The St. Lawrence system in Canada already demonstrates substantial deterioration that can be attributed to the human activity of the many millions of people who are situated on the shores of the Great Lakes and on the banks of the St. Lawrence and its tributaries. With so many people and so much activity, concern has been expressed that the alteration in the characteristics of the St. Lawrence will soon be reflected in changes in the Gulf of St. Lawrence all the way out to Cabot Strait.

With these kinds of examples increasingly apparent, the emphasis in resource management has more and more shifted from the technology of single activities, such as forestry or fisheries, to technologies for total resource management of entire watersheds or regions. It is no longer sufficient to make decisions about logging or about agriculture, or about industry, or about recreation, unless the whole development of a region is taken into consideration. To continue our erstwhile practices of minding just our own business can only lead to chaos and a real decline in our future potentials.

Holistic approaches to resource management have inevitably meant that economics and sociology and other social sciences are increasingly important contributors to resource management decisions. Instead of only considering what it would cost to install pollution-abatement facilities, it becomes necessary to ask, "What benefits will society receive from this industry, and what will it mean in terms of costs? Is there a possibility that in some instances we may prefer to have a short-term wealth in exchange for poverty in the long term? Alternatively, will the effects of pollution last so long and the costs be so large that we will not choose to have an industry even though it would give us a short-term standard of living that we desire?" These are difficult questions and they involve thoughtful judgements by every citizen. There can be little doubt that in Canada we are now at that stage of development when Canadians should vigorously express their opinions about the kind of country they wish their grandchildren to live in.

There is an element of urgency to thinking about matters of environmental quality. The rate of growth in Canadian population and industry has been extremely rapid over the last two decades. If every country in the world used resources at the same rate that Canada does, the world's supply of a great many of them would quickly be exhausted. We are fortunate in having a huge area that is well endowed with natural assets but our rate of growth has been such that we can already see that we are approaching the limits of our productive capacity. Virgin forests will soon be a thing of the past in Canada; and our production will depend not on the trees that we can "hunt" but on the trees that we can grow. Most of the good agricultural land in Canada (and some which is not so good) is already under cultivation. Some of our best agricultural land is now underneath suburban housing developments. It is accordingly the right time to make the crucial decisions that will influence our future. With the accelerating pace of events, it seems likely that we have already made many of the decisions that will adversely determine the lives of the next generations.

Many of these sentiments have been expressed with great vigour during the last five years and in some instances in almost hysterical terms. But they have increased public awareness and there is now considerable optimism that we will awake in time to take firm control over the future of our resources. There may be a substantial struggle in translating our good intentions into action. The present pattern of legislation concerning pollution in particular, and resource management in general, is muddled and confused. Municipal, provincial, and federal levels of government are involved in a bewildering complex of statutes that do not reflect an integrated appreciation of what man can do to his environment. In some ways it is funny when one hears anecdotes about mine managers who are visited by officials from fourteen different government departments, all of whom have some interest in pollution, or that the effects of the Bennett dam involve ten "key" statutes. But the afterthought to this kind of situation is far from funny. Obviously, the machineries of our modern bureaucracies are about as funny as the Keystone Kops would be if they were our local police force.

The passage of the Canada Water Act in 1970 (Appendix) may prove to be a large step in the direction of more effective control of water quality. The various regulations that this Act enables are currently being drafted and should provide a comprehensive set of minimum standards. If the Act is enforced with vigour, major improvements in pollution control should not be far behind.

There is urgent need for further legislative reform, and for enforcement of regulations, if we are to avoid the same mistakes that most other developed countries have already made in permitting the wholesale pollution of their lakes and streams. As we all know, legislative shifts and enforcement of regulations come about because politicians perceive that there is either a public desire for change or a wide public acceptance of imaginative leadership in planning for the future. It behooves us then to make known that we are anxious for reform and ready for leadership in providing for a clean future for Canada's lakes and streams.

Appendix

THE CANADA WATER ACT

An Act to provide for the management of the water resources of Canada including research and the planning and implementation of programs relating to the conservation, development and utilization of water resources.

[1969-70, c. 52]

WHEREAS the demands on the water resources of Canada are increasing rapidly and more knowledge is needed of the nature, extent and distribution of those resources, of the present and future demands thereon and of the means by which those demands may be met;

AND WHEREAS pollution of the water resources of Canada is a significant and rapidly increasing threat to the health, well-being and prosperity of the people of Canada and to the quality of the Canadian environment at large and as a result it has become a matter of urgent national concern that measures be taken to provide for water quality management in those areas of Canada most critically affected;

AND WHEREAS the Parliament of Canada is desirous that, in addition, comprehensive programs be undertaken by the Government of Canada and by the Government of Canada in cooperation with provincial governments, in accordance with the responsibilities of the federal government and each of the provincial governments in relation to water resources, for research and planning with respect to those resources and for their conservation, development and utilization to ensure their optimum use for the benefit of all Canadians;

Now, THEREFORE, Her Majesty, by and with the advice and consent of the Senate and House of Commons of Canada, enacts as follows:

SHORT TITLE

1. This Act may be cited as the *Canada Water Act.*

INTERPRETATION

2. (1) In this Act

"agency" means a water quality management agency the incorporation of which is procured or that is named pursuant to section 9 or 11;

"analyst" means an analyst designated pursuant to section 23;

"boundary waters" means the waters from main shore to main shore of the lakes and rivers and connecting waterways, or the portions thereof, along which the international boundary between the United States and Canada passes, including all bays, arms and inlets thereof, but not including tributary waters which in their natural channels would flow into such lakes, rivers and waterways, or waters flowing from such lakes, rivers and waterways, or the waters of rivers flowing across the boundary;

"federal agency" means a water quality management agency the incorporation of which is procured or that is named pursuant to section 11;

"federal waters" means waters under the exclusive legislative jurisdiction of the Parliament of Canada;

"inspector" means an inspector designated pursuant to section 23;

"inter-jurisdictional waters" means any waters, whether international, boundary or otherwise, that, whether wholly situated in a province or not, significantly affect the quantity or quality of waters outside such province;

"international waters" means waters of rivers that flow across the international boundary between the United States and Canada;

"Minister" means the Minister of Energy, Mines and Resources;

"prescribed" means prescribed by regulation;

"waste" means any substance that, if added to any waters, would degrade or alter or form part of a process of degradation or alteration of the quality of those waters to an extent that is detrimental to their use by man or by any animal, fish or plant that is useful to man, and includes any water that contains a substance in such a quantity or concentration, or that has been so treated, processed or changed, by heat or other means, from a natural state that it would, if added to any waters, degrade or alter or form part of a process of degradation or alteration of the quality of those waters to an extent that is detrimental to their use by man or by any animal, fish or plant that is useful to man;

"water quality management" means any aspect of water resource management that relates to restoring, maintaining or improving the quality of water;

"water resource management" means the conservation, development and utilization of water resources, and includes, with respect thereto, research, data collection and the maintaining of inventories, planning and the implementation of plans, and the control and regulation of water quantity and quality.

(2) Without limiting the generality of the definition of the term "waste" in this Act,

(a) any substance or any substance that is part of a class of substances prescribed pursuant to subparagraph 16(1)(a)(i),

(b) any water that contains any substance or any substance that is part of a class of substances in a quantity or concentration that is equal to or in excess of a quantity or concentration prescribed in respect of that substance or class of substances pursuant to subparagraph 16(1)(a)(ii), and

(c) any water that has been subjected to a treatment, process or change prescribed pursuant to subparagraph 16(1)(a)(iii),

shall, for the purposes of this Act, be deemed to be waste.

(3) This Act is binding on Her Majesty in right of Canada or a province and any agent thereof.

PART I

COMPREHENSIVE WATER RESOURCE MANAGEMENT

Federal-Provincial Arrangements

3. For the purpose of facilitating the formulation of policies and programs with respect to the water resources of Canada and to ensure the optimum use of those resources for the benefit of all Canadians, having regard to the distinctive geography of Canada and the character of water as a natural resource, the Minister may, with the approval of the Governor in Council, enter into an arrangement with one or more provincial governments to establish, on a national, provincial, regional or lake or river-basin basis, intergovernmental committees or other bodies.

(a) to maintain continuing consultation on water resource matters and to advise on priorities for research, planning, conservation, development and utilization relating thereto;

(b) to advise on the formulation of water policies and programs; and

(c) to facilitate the coordination and implementation of water policies and programs.

Comprehensive Water Resource Management Programs

4. Subject to this Act, the Minister may, with the approval of the Governor in Council, with respect to any waters where there is a significant national interest in the water resource management thereof, from time to time enter into agreements with one or more provincial governments having an interest in the water resource management of those waters, providing for programs to

(a) establish and maintain an inventory of those waters,

(b) collect, process and provide data on the quality, quantity, distribution and use of those waters,

(c) conduct research in connection with any aspect of those waters or provide for the conduct of any such research by or in cooperation with any government, institution or person,

(d) formulate comprehensive water resource management plans, including detailed estimates of the cost of implementation of those plans and of revenues and other benefits likely to be realized from the implementation thereof, based upon an examination of the full range of reasonable alternatives and taking into account views expressed at public hearings and otherwise by persons likely to be affected by implementation of the plans,

(e) design projects for the efficient conservation, development and utilization of those waters, and

(f) implement any plans or projects referred to in paragraphs *(d)* and *(e)*, and establishing or naming joint commissions, boards or other bodies empowered to direct, supervise and coordinate such programs.

5. (1) Subject to subsection (2), the Minister shall, with the approval of the Governor in Council, undertake directly,

(a) with respect to any federal waters, any program described in any of paragraphs 4*(a)* to *(e)* and the implementation of any program described in paragraph 4*(d)* or *(e)*;

(b) with respect to any inter-jurisdictional waters where there is a significant national interest in the water resource management thereof, any program described in paragraph 4*(d)* or *(e)*; and

(c) with respect to any international or boundary waters where there is a significant national interest in the water resource management thereof, any program described in paragraph 4*(d)* or *(e)* and the implementation of any such program.

(2) The Governor in Council shall not approve the undertaking by the Minister of any program pursuant to paragraph 1*(b)* or *(c)* unless he is satisfied that all reasonable efforts have been made by the Minister to reach an agreement under section 4 with the one or more provincial governments having an interest in the water resource management of the waters in question, and that those efforts have failed.

(3) In undertaking any programs pursuant to subsection (1), the Minister shall take into account any priorities for development recommended pursuant to paragraph 3*(a)*.

6. The Minister may conduct research, collect data and establish inventories respecting any aspect of water resource management or the management of any specific water resources or provide for the conduct of any such research, data collection or inventory establishment by or in cooperation with any government, institution or person.

7. (1) Where, pursuant to section 4, the Minister enters into an agreement with one or more provincial governments, the agreement shall, where applicable, in respect of each program that is the subject of such agreement, specify

(a) the respective parts of the program to be undertaken by the Minister and the provincial government or governments that are parties to the agreement and the times at which and the manner in which such parts of the program are to be carried out;

(b) the proportions of the cost of the respective parts of the program that are to be paid by the Minister and the provincial government or governments and the times at which amounts representing such proportions are to be paid;

(c) the labour, land and materials that are to be supplied in relation to the respective parts of the program by the Minister and the provincial government or governments;

(d) the proportions in which any compensation awarded or agreed to be paid to any body or person suffering loss as a result of the program is to be paid by the Minister and the provincial government or governments;

(e) the amount of any loan or grant, constituting part or all of the cost of the program that is to be paid by the Minister, that is to be made or paid by the Minister to the provincial government or governments, and the manner in which the terms and conditions of the loan or grant are to be determined;

(f) the authority or authorities, whether an agent or agents of Her Majesty in right of Canada or a province or otherwise as may be agreed to be appropriate, that will be responsible for the undertaking, operation or maintenance of projects that form part of the program;

(g) the respective proportions of the revenues from the program that are to be paid to the Minister and the provincial government or governments; and

(h) the terms and conditions relating to the undertaking, operation and maintenance of the program.

(2) An agreement entered into pursuant to section 4 shall, where applicable, in respect of the joint commission, board or other body thereby established or named, provide for

(a) the constitution thereof, the members thereof that are to be appointed by the Minister and the provincial government or governments that are parties to the agreement and the terms and conditions of such appointments;

(b) the staff thereof that is to be supplied by the Minister and the provincial government or governments;

(c) the duties of the body and the powers that it may exercise in directing, supervising and coordinating the program;

(d) the keeping of accounts and records by the body;

(e) the annual submission by the body to the Minister and the provincial government or governments of operating and capital budgets in connection with the programs directed, supervised and coordinated by the body for the next following fiscal year for approval by the Governor in Council and such persons on behalf of the provincial government or governments as are designated in the agreement; and

(f) the submission by the body to the Minister and the provincial government or governments, within three months after the termination of each fiscal year, of an annual report containing such information as is specified in the agreement.

PART II

WATER QUALITY MANAGEMENT

Pollution of Waters

8. Except in quantities and under conditions prescribed with respect to waste disposal in the water quality management area in question, including the payment of any effluent discharge fee prescribed therefor, no person shall deposit or permit the deposit of waste of any type in any waters comprising a water quality management area designated pursuant to section 9 or 11, or in any place under any conditions where such waste or any other waste that results from the deposit of such waste may enter any such waters.

Federal-Provincial Water Quality Management

9. In the case of

(a) any waters, other than federal waters, the water quality management of which has become a matter of urgent national concern, or

(b) any federal waters,

the Minister may, with the approval of the Governor in Council, from time to time enter into agreements with one or more provincial governments having an interest in the water quality management thereof, designating those waters as a water quality management area, providing for water quality management programs in respect thereof and authorizing the Minister, jointly with such one or more provincial governments, to procure the incorporation of a corporation without share capital, or to name an existing corporation that is an agent of Her Majesty in right of Canada or a province or that performs any function or duty on behalf of the Government of Canada or the government of a province, as a water quality management agency to plan, initiate and carry out, in conjunction with the Minister and such provincial government or governments, programs described in section 13 in respect of those waters.

10. (1) Where, pursuant to section 9, the Minister enters into an agreement with one or more provincial governments, the agreement shall, where applicable, in respect of each water quality management program that is the subject of such agreement, specify

(a) the responsibilities of the Minister and the provincial government or governments that are parties to the agreement and the times at which and the manner in which such responsibilities are to be undertaken;

(b) the proportions of the capital cost of the respective parts of the program that are to be paid by the Minister and the provincial government or governments and the times at which amounts representing such proportions are to be paid;

(c) the loans or contributions in respect of the cost of incorporation and operating expenses of the agency and the loans in respect of capital costs incurred by the agency that is to undertake the program, that are to be made or paid by the Minister and the provincial government or governments and the times at which such loans or contributions are to be made or paid;

(d) the labour, land and materials that are to be supplied by the Minister and the provincial government or governments to the agency that is to undertake the program;

(e) the proportions in which any compensation awarded or agreed to be paid to any body or person suffering loss as a result of the program is to be paid by the Minister and the provincial government or governments; and

(f) the terms and conditions relating to the undertaking, operation and maintenance of the program by the agency,

and each such agreement shall provide that it may be terminated, on six months written notice by any party to the agreement to all other parties thereto or on such lesser notice as may be agreed upon by all such parties, and that upon the expiration of the time fixed by such notice for the termination of the agreement any agency incorporated thereunder shall be wound up.

(2) An agreement entered into pursuant to section 9 shall, in respect of the agency the incorporation of which is thereby authorized to be procured, if any, provide for

(a) the proposed corporate name of the agency;

(b) the place within the water quality management area designated in the agreement where the head office of the agency is to be situated;

(c) the members thereof that are to be appointed by the Minister and the provincial government or governments that are parties to the agreement and the terms and conditions of such appointments;

(d) the proposed by-laws of the agency; and

(e) the matters set out in paragraphs 7(2)*(b)* and 7(2)*(d)* to *(f)*.

Federal Water Quality Management

11. (1) Where, in the case of any inter-jurisdictional waters, the water quality management of those waters has become a matter of urgent national concern, and either

(a) the Governor in Council is satisfied that all reasonable efforts have been made by the Minister to reach an agreement under section 9 with the one or more provincial governments having an interest in the water quality management thereof, and that those efforts have failed, or

(b) although an agreement was reached under section 9 in respect thereof and an agency was incorporated or named thereunder, the Minister and the appropriate minister of each provincial government that was a party to the agreement disagreed with the recommendations of the agency with respect to water quality standards for those inter-jurisdictional waters and were unable to agree on a joint recommendation with respect thereto and as a result thereof the agreement under section 9 was terminated,

the Governor in Council may, on the recommendation of the Minister, designate such waters as a water quality management area and authorize the Minister to procure the incorporation of a corporation without share capital under Part II of the *Canada Corporations Act,* or to name an existing corporation that is an agent of Her Majesty in right of Canada or that performs any function or duty on behalf of the Government of Canada, as a water quality management agency to plan, initiate and carry out programs described in section 13 in respect of those waters.

(2) The Governor in Council may, on the recommendation of the Minister, designate any federal waters as a water quality management area and authorize the Minister to procure the incorporation of a corporation without share capital under Part II of the *Canada Corporations Act,* or to name an existing corporation that is an agent of Her Majesty in right of Canada or that performs any function or duty on behalf of the Government of Canada, as a water quality management agency to plan, initiate and carry out programs described in section 13 in respect of those waters.

(3) In procuring the incorporation of a federal agency pursuant to subsection (1) or (2), the Minister shall, with the approval of the Governor in Council, provide for

(a) the appointment of the members thereof by the Governor in Council for such terms and under such conditions as the Governor in Council deems suitable;

(b) the staff thereof that is to be supplied by the Minister; and

(c) the terms and conditions under which and the remuneration at which any staff may be appointed by the federal agency.

(4) The Minister may give directions to any federal agency with respect to the implementation of any water quality management program and, in so doing, he shall take into account any priorities for development recommended pursuant to paragraph 3*(a)*.

12. (1) A federal agency is for all purposes an agent of Her Majesty and its powers may be exercised only as an agent of Her Majesty.

(2) A federal agency may, on behalf of Her Majesty, enter into contracts in the name of Her Majesty or in its name.

(3) Property acquired by a federal agency is the property of Her Majesty and title thereto may be vested in the name of Her Majesty or in its name.

(4) Actions, suits or other legal proceedings in respect of any right or obligation acquired or incurred by a federal agency on behalf of Her Majesty, whether in its name or in the name of Her Majesty, may be brought or taken by or against that agency in the name of the agency in any court that would have jurisdiction if it were not an agent of Her Majesty.

Water Quality Management Agencies

13. (1) The objects of each water quality management agency shall be to plan, initiate and carry out programs to restore, preserve and enhance the water quality level in the water quality management area for which the agency is incorporated or named and in carrying out those objects, subject to any agreement under section 9 relating to such water quality management area or to any direction of the Minister to a federal agency, the agency may, after taking into account views expressed to it, at public hearings and otherwise, by persons likely to have an interest therein, in respect of the waters comprising the water quality management area for which it is incorporated or named,

(a) ascertain the nature and quantity of waste present therein and the water quality level;

(b) undertake studies that enable forecasts to be made of the amounts and kinds of waste that are likely to be added to those waters in the future;

(c) develop and recommend to the Minister and, in the case of an agency other than a federal agency, to the appropriate minister of each provincial government that is a party to the agreement relating to the water quality management area, a water quality management plan including

(i) recommendations as to water quality standards for those waters or any part thereof and the times at which those standards should be attained,

(ii) recommendations, based upon the water quality standards recommended pursuant to subparagraph (i), as to the quantities and types of waste, if any, that may be deposited in those waters and the conditions under which any such waste may be deposited,

(iii) recommendations as to the treatment that may be required for any waste that is or may be deposited in those waters and the type of treatment facilities necessary to achieve the water quality standards recommended pursuant to subparagraph (i),

(iv) recommendations as to appropriate effluent discharge fees to be paid by persons for the deposit of waste in those waters and the time or times at which and the manner in which such fees should be paid,

(v) recommendations as to appropriate waste treatment and waste sample analysis charges to be levied by the agency for the treatment of waste at any waste treatment facility that is operated and maintained by it or for the analysis of any waste sample by it,

(vi) detailed estimates of the cost of implementation of the plan and of revenues and other benefits likely to be realized from the implementation thereof, and

(vii) estimates of the time within which the agency would become financially self-sustaining.

(2) Where an agency recommends a water quality management plan to the Minister, it shall forthwith cause the plan to be published in the *Canada Gazette* and shall publish a concise summary of the plan in a newspaper of general circulation in the area affected by the plan at least once a week for a period of four weeks; no such plan shall be approved until the expiration of seven clear days after the publication last so required.

(3) Where a water quality management plan recommended by an agency in respect of the waters comprising the water quality management area for which it is incorporated or named has been approved by the Minister and, in the case of an agency other than a federal agency, by the appropriate minister of each provincial government that is a party to the agreement relating to those waters,

the agency may, in order to implement the water quality management plan,

(a) design, construct, operate and maintain waste treatment facilities and undertake the treatment of waste delivered to such facilities;

(b) undertake the collection of any charges prescribed for waste treatment at any waste treatment facility that is operated and maintained by it and for waste sample analysis carried out by it;

(c) undertake the collection of effluent discharge fees prescribed to be payable by any person for the deposit of waste in those waters;

(d) monitor, on a regular basis, water quality levels;

(e) provide facilities for the analysis of samples of waste and collect and provide data respecting the quantity and quality of waste and the effects thereof on those waters;

(f) regularly inspect any waste treatment facilities within the water quality management area for which it is incorporated or named, whether publicly or privately owned;

(g) publish or otherwise distribute such information as may be required under this Act; and

(h) do such other things as are necessary to achieve effective water quality management of those waters.

(4) Except with respect to loans authorized to be made by it by the Minister or a provincial government as described in paragraph 10(1) *(c)*, an agency does not have power to borrow moneys, to issue securities or to guarantee the payment of any debt or obligation of any person.

14. (1) The members of an agency who are appointed by the Minister or by the Governor in Council and who are not employees in the public service of Canada shall be paid by the agency such remuneration as is authorized by the Governor in Council.

(2) The Minister may provide any agency with such officers and employees as may be necessary for the proper functioning of the agency, and may provide any such agency with such professional or technical assistance for temporary periods or for specific work as the agency may request.

(3) Subject to the agreement under which the incorporation of an agency was authorized to be procured or to any matter provided for under subsection 11(3) or any direction of the Minister under subsection 11(4) in respect of a federal agency, the agency may employ such officers and employees and such consultants and advisers as it considers necessary to enable it to carry out its objects and fix the terms and conditions of their employment and their remuneration, which shall be paid by the agency.

15. (1) Each agency shall maintain under the name of the agency, in a chartered bank, an account to which shall be deposited

(a) all amounts collected by the agency as or on account of charges levied for treatment of waste, the analysis of samples of waste or for deposit of waste in the waters comprising the water quality management area for which the agency is incorporated or named,

(b) contributions paid or loans made to the agency by the Government of Canada or the government of a province in respect of the cost of incorporation of the agency, in respect of its operating expenses or in respect of capital costs incurred by it, and

(c) interest received by the agency on securities purchased, acquired and held by it pursuant to subsection (2),

and out of which shall be paid all expenditures incurred by the agency in its operations and all repayments of loans made to the agency and payments of interest thereon.

(2) An agency may from time to time, out of any surplus funds standing to its credit in an account established pursuant to subsection (1), purchase, acquire and hold

(a) in the case of a federal agency, any securities of or guaranteed by the Government of Canada; and

(b) in the case of any other agency, any securities of or guaranteed by the Government of Canada, or of or guaranteed by the government of any province that is a party to the agreement pursuant to which the agency was authorized to be incorporated or was named.

(3) An agency may sell any securities purchased, acquired and held pursuant to subsection (2) and the proceeds of sale shall be deposited to the credit of the agency in the account established in respect of the agency under subsection (1).

Regulations

16. (1) The Governor in Council may make regulations

(a) prescribing

(i) substances and classes of substances,

(ii) quantities or concentrations of substances and classes of substances in water, and

(iii) treatments, processes and changes of water

for the purpose of subsection 2(2);

(b) prescribing the procedure to be followed by each agency in determining its recommendations as to charges that may be levied by it for treatment of waste at any waste treatment facility that is operated and maintained by the agency;

(c) prescribing the procedure to be followed by each agency in determining its recommendations as to water quality standards for waters comprising the water quality management area for which it is incorporated or named;

(d) prescribing the criteria, which shall be related to estimates of the cost of appropriate treatment of waste expected to be deposited, to be used by each agency in determining its recommendations as to effluent discharge fees to be paid by persons for the deposit of waste in waters comprising the water quality management area for which it is incorporated or named and the time or times at which and the manner in which such fees should be paid;

(e) requiring persons who deposit waste in any waters comprising a water quality management area to maintain books and records necessary for the proper enforcement of this Act and the regulations;

(f) requiring persons who have deposited waste in contravention of section 8 to report such deposit to the agency incorporated or named for the water quality management area in which the deposit is made and providing for the manner in which and the time within which such report is to be made;

(g) requiring persons who deposit waste in any waters comprising a water quality management area to submit test portions of such waste to the agency incorporated or named in respect of the area;

(h) respecting the method of analysis by each agency of test portions of waste submitted to it;

(i) respecting the powers and duties of inspectors and analysts, the taking of samples and the making of analyses for the purposes of this Act; and

(j) generally, for carrying out the purposes and provisions of this Act.

(2) Subject to subsection (3), the Governor in Council may make regulations prescribing, with respect to each water quality management area,

(a) the quantities, if any, of waste of any type that for the purposes of section 8, may be deposited in the waters comprising such area and the conditions under which any such waste may be deposited;

(b) the charges to be paid by any person to the agency incorporated or named in respect thereof

    (i) for treatment of waste by the agency at a waste treatment facility that is operated and maintained by it, and

    (ii) for analysis of waste samples by the agency,

and the persons by whom such charges are payable and the time or times at which and the manner in which such charges shall be paid;

(c) water quality standards for the waters comprising such area; and

(d) the effluent discharge fees, if any, to be paid by any person to the agency incorporated or named in respect thereof for the deposit of waste in the waters comprising such area and the persons by whom such fees are payable and the time or times at which and the manner in which such fees shall be paid.

(3) No regulation that is made by the Governor in Council under subsection (2) with respect to a water quality management area for which an agency is incorporated or named under an agreement entered into pursuant to section 9 is of any force or effect unless

(a) it is made on the recommendation of the agency, or

(b) where the Minister and the appropriate minister of each provincial government that is a party to the agreement disagree with the recommendations of the agency and jointly make a different recommendation, it is made on such joint recommendation.

PART III

NUTRIENTS

Interpretation

17. In this Part and Part IV,

"cleaning agent" means any laundry detergent, dishwashing compound, household cleaner, metal cleaner, degreasing compound, commercial cleaner, industrial cleaner, phosphate compound or other substance intended to be used for cleaning purposes;

"nutrient" means any substance or combination of substances that, if added to any waters in sufficient quantities, provides nourishment that promotes the growth of aquatic vegetation in those waters to such densities as to

(a) interfere with their use by man or by any animal, fish or plant that is useful to man, or

(b) degrade or alter or form part of a process of degradation or alteration of the quality of those waters to an extent that is detrimental to their use by man or by any animal, fish or plant that is useful to man;

"water conditioner" means any water softening chemical, anti-scale chemical, corrosion inhibiter or other substance intended to be used to treat water.

Use of Nutrients

18. No person shall manufacture for use or sale in Canada or import into Canada any cleaning agent or water conditioner that contains a prescribed nutrient in a concentration that is greater than the prescribed maximum permissible concentration of that nutrient in that cleaning agent or water conditioner.

Regulations

19. The Governor in Council may make regulations

(a) prescribing, for the purpose of section 18,
  (i) nutrients, and
  (ii) the maximum permissible concentration, if any, of any prescribed nutrient in any cleaning agent or water conditioner;
(b) respecting the manner in which the concentration of any prescribed nutrient in a cleaning agent or water conditioner shall be determined; and
(c) requiring persons who manufacture in Canada or import into Canada any cleaning agent or water conditioner
  (i) to maintain books and records necessary for the proper enforcement of this Part and regulations made under this section, and
  (ii) to submit samples of such cleaning agent or water conditioner to the Minister.

Seizure

20. (1) An inspector may at any reasonable time seize any cleaning agent or water conditioner that he reasonably believes has been manufactured in Canada or imported into Canada in violation of section 18.

(2) Any cleaning agent or water conditioner seized under this Act by an inspector may at the option of an inspector be kept or stored in the building or place where it was seized or may be removed to any other proper place by or at the direction of an inspector.

(3) Except with the authority of an inspector, no person shall remove, alter or interfere in any way with any cleaning agent or water conditioner seized under this Act by an inspector; but an inspector shall, at the request of a person from whom any cleaning agent or water conditioner was so seized, furnish a sample thereof to that person for analysis.

21. (1) Where any cleaning agent or water conditioner has been seized under this Act, any person may, within two months after the date of such seizure, upon prior notice having been given in accordance with subsection (2) to the Minister by registered mail addressed to him at Ottawa, apply to a magistrate within whose territorial jurisdiction the seizure was made for an order of restoration under subsection (3).

(2) The notice referred to in subsection (1) shall be mailed at least fifteen clear days prior to the day on which the application is to be made to the magistrate and shall specify

(a) the magistrate to whom the application is to be made;
(b) the place where and the time when the application is to be heard;
(c) the cleaning agent or water conditioner in respect of which the application is to be made; and
(d) the evidence upon which the applicant intends to rely to establish that he is entitled to possession of the cleaning agent or water conditioner in respect of which the application is to be made.

(3) Subject to section 22, where, upon the hearing of an application made under subsection (1), the magistrate is satisfied

(a) that the applicant is otherwise entitled to possession of the cleaning agent or water conditioner seized, and
(b) that the cleaning agent or water conditioner seized is not and will not be required as evidence in any proceedings in respect of an offence under this Act,

he shall order that the cleaning agent or water conditioner seized be restored forthwith to the applicant, and where the magistrate is satisfied that the applicant is otherwise entitled to possession of the cleaning agent or water conditioner seized but is not satisfied as to the matters mentioned in paragraph *(b)*, he shall order that the cleaning agent or water conditioner seized be restored to the applicant

(c) upon the expiration of four months from the date of such seizure if no proceedings in respect of a violation of section 18 have been commenced before that time, or
(d) upon the final conclusion of any such proceedings in any other case.

(4) Where no application has been made under subsection (1) for the restoration of any cleaning agent or water conditioner seized under this Act within two months from the date of such seizure, or an application therefor has been made but upon the hearing thereof no order of restoration is made, the cleaning agent or water conditioner so seized shall be delivered to the Minister who may make such disposition thereof as he thinks fit.

22. (1) Where a person is convicted of an offence under subsection 28(1), any cleaning agent or water conditioner seized under this Act by means of or in respect of which the offence was committed is thereupon forfeited to Her Majesty and shall be disposed of as the Minister directs.

(2) Where an inspector has seized any cleaning agent or water conditioner under this Act and the owner thereof or the person in whose possession it was at the time of seizure consents in writing to the destruction thereof, the cleaning agent or water conditioner is thereupon forfeited to Her Majesty and shall be disposed of as the Minister directs.

PART IV

GENERAL

Inspectors and Analysts

23. The Minister may designate any qualified person as an inspector or analyst for the purposes of this Act but where a qualified officer of any other department or agency of the Government of Canada carries out similar duties for the purposes of another Act the Minister shall designate such officer whenever possible.

24. (1) An inspector may at any reasonable time
(a) enter any area, place, premises, vessel or vehicle, other than a private dwelling place or any part of any such area, place, premises, vessel or vehicle that is designed to be used and is being used as a permanent or temporary private dwelling place, in which he reasonably believes
    (i) there is being or has been carried out any manufacturing or other process that may result in or has resulted in waste, or
    (ii) there is any waste
that may be or has been added to any waters that have been designated as a water quality management area pursuant to section 9 or 11;
(b) enter any area, place, premises, vessel or vehicle, other than a private dwelling place or any part of any such area, place, premises, vessel or vehicle that is designed to be used and is being used as a permanent or temporary private dwelling place, in which he reasonably believes
    (i) any cleaning agent or water conditioner is being manufactured, or
    (ii) there is any cleaning agent or water conditioner that has been manufactured in Canada or imported into Canada in violation of section 18;
(c) examine any waste, cleaning agent or water conditioner found therein in bulk or open any container found therein that he has reason to believe contains any waste, cleaning agent or water conditioner and take samples thereof; and

(d) require any person in such area, place, premises, vehicle or vessel to produce for inspection or for the purpose of obtaining copies thereof or extracts therefrom, any books or other documents or papers concerning any matter relevant to the administration of this Act or the regulations.

(2) An inspector shall be furnished with a certificate of his designation as an inspector and on entering any area, place, premises, vehicle or vessel referred to in subsection (1) shall, if so required, produce the certificate to the person in charge thereof.

(3) The owner or person in charge of any area, place, premises, vehicle or vessel referred to in subsection (1) and every person found therein shall give an inspector all reasonable assistance in his power to enable the inspector to carry out his duties and functions under this Act and the regulations and shall furnish him with such information with respect to the administration of this Act and the regulations as he may reasonably require.

25. (1) No person shall obstruct or hinder an inspector in the carrying out of his duties or functions under this Act or the regulations.

(2) No persons shall knowingly make a false or misleading statement, either verbally or in writing, to an inspector or other person engaged in carrying out his duties or functions under this Act or the regulations.

Advisory Committees

26. (1) The Minister may establish and appoint the members of such advisory committees as he considers desirable for the purpose of advising and assisting him in carrying out the purposes and provisions of this Act.

(2) Each member of an advisory committee is entitled to be paid reasonable travelling and other expenses while absent from his ordinary place of residence in the course of his duties as such a member and may, with the approval of the Minister, be paid such amount as is fixed by the Governor in Council for each day he attends any meeting of the committee or for each day during which he performs, with the approval of the committee, any duties on behalf of the committee in addition to his ordinary duties as a member thereof.

Public Information Program

27. The Minister may, either directly or in cooperation with any government, institution or person, publish or otherwise distribute or arrange for the publication or distribution of such information as he deems necessary to inform the public respecting any aspect of the conservation, development or utilization of the water resources of Canada.

Offences

28. (1) An person who violates section 8 or 18 is liable on summary conviction to a fine not exceeding five thousand dollars for each offence.

(2) Where an offence under subsection (1) is committed on more than one day or is continued for more than one day, it shall be deemed to be a separate offence for each day on which the offence is committed or continued.

29. Any person who violates subsection 20(3) or section 25 or any regulation made under paragraph 16(1)*(e)*, *(f)* or *(g)* or paragraph 19*(c)* is guilty of an offence punishable on summary conviction.

30. Where a person is convicted of an offence under this Act, the court may, in addition to any punishment it may impose, order that person to refrain from any further violation of the provision of the Act or regulations for the violation of which he has been convicted or to cease to carry on any activity specified in the order the carrying on of which, in the opinion of the court, will or is likely to result in any further violation thereof.

31. In a prosecution for an offence under this Act, it is sufficient proof of the offence to establish that it was committed by an employee or agent of the accused whether or not the employee or agent is identified or has been prosecuted for the offence, unless the accused establishes that the offence was committed without his knowledge or consent and that he exercised all due diligence to prevent its commission.

32. Proceedings in respect of an offence under this Act may be instituted at any time within two years after the time when the subject-matter of the proceedings arose.

33. Any complaint or information in respect of an offence under this Act may be heard, tried or determined by a court if the accused is resident or carrying on business within the territorial jurisdiction of that court although the matter of the complaint or information did not arise in that territorial jurisdiction.

34. (1) Notwithstanding that a prosecution has been instituted in respect of an offence under subsection 28(1) the Attorney General of

Canada may commence and maintain proceedings to enjoin any violation of section 8.

(2) No civil remedy for any act or omission is suspended or affected by reason that the act or omission is an offence under this Act.

Evidence

35. (1) Subject to this section, a certificate of an analyst stating that he has analyzed or examined a sample submitted to him by an inspector and stating the result of his analysis or examination is admissible in evidence in any prosecution for a violation of this Act and in the absence of evidence to the contrary is proof of the statements contained in the certificate without proof of the signature or the official character of the person appearing to have signed the certificate.

(2) The party against whom a certificate of an analyst is produced pursuant to subsection (1) may, with leave of the court, require the attendance of the analyst for the purposes of cross-examination.

(3) No certificate shall be received in evidence pursuant to subsection (1) unless the party intending to produce it has given to the party against whom it is intended to be produced reasonable notice of such intention together with a copy of the certificate.

Report to Parliament

36. The Minister shall, as soon as possible after the end of each fiscal year, prepare a report on the operations under this Act for that year, and the Minister shall cause such report to be laid before Parliament forthwith upon the completion thereof, or, if Parliament is not then sitting, on any of the first fifteen days next thereafter that Parliament is sitting.

Financial

37. All expenditures by the Minister for the purposes of this Act shall be paid out of moneys appropriated by Parliament therefor.

38. Subject to section 37, the Minister may, with the approval of the Governor in Council,

(a) make loans or pay contributions to any agency in respect of the cost of incorporating the agency or in respect of its operating expenses or make loans to any agency in respect of capital costs incurred by it; and

(b) in accordance with an agreement entered into under section 4, make loans or pay grants to the government of any province to meet any part of the portion of the cost of programs undertaken pursuant to such an agreement that is to be paid by the Minister.

APPLICATION

39. Section 8 is not applicable in respect of a water quality management area designated pursuant to section 9 or 11 until a proclamation has been issued declaring it to be applicable in respect of that area.

The Canada Water Act, Chapter 52, 18-19 Elizabeth II, 1970. Reproduced by permission of Information Canada.

REFERENCES

SOME USEFUL REFERENCE BOOKS ON POLLUTION

There are many recent books on environmental problems. Listed below are some that are commonly used. This list is not complete, but provides enough to get a good look into the literature.

COMPREHENSIVE TECHNICAL BOOKS ON MAN'S EFFECT ON HIS ENVIRONMENT

DETWYLER, THOMAS R. (ed.)
  1971. Man's impact on environment. New York: McGraw-Hill. (A collection of scientific articles, many of a review nature, on a variety of environmental problems, including nine on different aspects of water pollution.)

GOODMAN, GORDON T.; EDWARDS, R. W.; and LAMBERT, J. M. (eds.)
  1965. Ecology and the industrial society. New York: John Wiley & Sons. (A publication based on a symposium of the British Ecological Society. Authoritative review papers on various pollution problems, many related to freshwater pollution.)

THE INSTITUTE OF ECOLOGY
  1973. Man in the living environment: report of the Workshop on Global Ecological Problems. Published by the Institute of Ecology, Washington, D.C. (An overview of world resource and pollution problems that probes into the possible crises for the future.)

MURDOCH, WILLIAM W. (ed.)

  1971. Environment: resources, pollution and society. Stamford,
    Connecticut: Sinauer. (A book to which many scientists
    contributed, that reviews the full spectrum of
    contemporary environmental problems.)

TECHNICAL BOOKS ON WATER POLLUTION

BARRY, P. J. (ed.)

  1972. Some aspects of the release of radioactivity and heat to
    the environment from nuclear reactors in Canada.
    Atomic Energy of Canada Limited, AECL-4156.

HYNES, H. B. N.

  1960. The biology of polluted waters. Liverpool: Liverpool
    University Press. (A widely used and excellent summary
    of what has been discovered in the study of polluted
    lakes and streams.)

JONES, J. R. E.

  1964. Fish and river pollution. London: Butterworth. (An
    excellent technical book on the effects of pollution on
    fish that inhabit streams.)

NATIONAL ACADEMY OF SCIENCES

  1969. Eutrophication: causes, consequences, correctives.
    Washington, D.C.: National Academy of Sciences. (A
    publication based on a symposium sponsored by the
    National Academy of Sciences. This book gives a fairly
    complete summary of what was known at the time about
    eutrophication of lakes and streams.)

U.S. DEPARTMENT OF THE INTERIOR, FEDERAL WATER POLLUTION
CONTROL ADMINISTRATION

  1970. Industrial waste guide on logging practices. U.S. Dept.
    Int., Northwest Region, Portland, Oregon.

WARREN, CHARLES E.

  1971. Biology and water pollution control. Toronto: W. B.
    Saunders. (A recent and excellent text for university
    courses on biology of polluted waters. A good base for
    professional training.)

BOOKS ON CANADIAN POLLUTION PROBLEMS

AULD, D. A. L. (ed.)
   1972. Economic thinking and pollution problems. Toronto:
   University of Toronto Press. (A collection of articles on
   economics of pollution problems. Many of the articles
   are by Canadians and about Canada. The appendices
   contain the text of the 1970 Canada Water Act.)

BATES, DAVID V.
   1972. A citizen's guide to air pollution. Montreal:
   McGill-Queen's University Press. (Second in the series
   sponsored by the Canadian Society of Zoologists. A
   concise statement of the nature of air-pollution problems
   with Canadian examples.)

BRINKHURST, RALPH O., and CHANT, DONALD A.
   1971. This good, good earth: our fight for survival. Toronto:
   Macmillan of Canada. (A readable account of
   environmental problems with particular reference to
   Canada.)

DUNBAR, M. J.
   1971. Environment and good sense: an introduction to
   environmental damage and control in Canada. Montreal:
   McGill-Queen's University Press. (First in the series
   sponsored by the Canadian Society of Zoologists. A
   broad review of environmental problems in Canada.)

EFFORD, I. E., and SMITH, BARBARA M. (eds.)
   1972. Energy and the environment. H. R. MacMillan Lectures,
   University of British Columbia, Vancouver. (A series of
   public lectures in 1971, sponsored by the Institute of
   Animal Resource Ecology, University of British
   Columbia, concerning energy demand.)

LAWRENCE, R. D.
   1969. The poison makers. Thos. Nelson & Sons (Canada) Ltd.

SCIENTIFIC PAPERS ON POLLUTION IN CANADA

A large proportion of the information on pollution in Canadian lakes
and streams is contained in unpublished reports in the files of various

government agencies. The material that is published is largely of a scientific nature, and indicates that Canada has a substantial number of pollution experts. The list below is drawn mostly from biological journals and is chiefly concerned with documenting the effects of various pollutants. There is also considerable literature in scientific and trade journals concerning pollution-treatment techniques and procedures. Information on local pollution problems can usually be obtained from provincial or federal government departments.

A comprehensive listing of pollution-research work in Canada, and of the people who do it, is contained in the volume *Inventory of Pollution-Relevant Research in Canada,* published in 1972 by the National Research Council of Canada (NRC No. 12678).

ALDERDICE, D. F., and WORTHINGTON, M. E.
> 1959. Toxicity of a DDT forest spray to young salmon. Canadian Fish Culturist 24:41–48.

ANDERSON, D. V. (ed.)
> 1969. The Great Lakes as an environment. Lecture Series, Great Lakes Institute Report PR39. University of Toronto.

ANON.
> 1970. Mercury in Lake St. Clair. Science News 97(16): 388.

BEAMISH, R. J., and HARVEY, H. H.
> 1972. Acidification of the LaCloche mountain lakes, Ontario, and resulting fish mortalities. Journal of the Fisheries Research Board of Canada 29:1131–43.

BEETON, A. M.
> 1961. Environmental changes in Lake Erie. Transactions of the American Fisheries Society 90(2):153–59.
> 1965. Eutrophication of the St. Lawrence Great Lakes. Limnology and Oceanography 10(2):240–54.

BERG, O. W.; DIOSADY, P. L.; and REES, G. A. V.
> 1968. Isolation of Dowtherm A in carp. Proceedings, Canadian Symposium on Water Pollution Research, 3:1–11.

CANADA DEPARTMENT OF ENVIRONMENT, FISHERIES SERVICE
> 1971. Summaries of fisheries research on the pollution problem. Research Subcommittee, Fisheries Development Council, Department of Environment. Vancouver.

CANADA DEPARTMENT OF NATIONAL HEALTH AND WELFARE, PUBLIC HEALTH ENGINEERING DIVISION
   1961. Survey report, effects of pollution on the Saint John River, New Brunswick and Maine, 1960. Department of National Health and Welfare, Ottawa.

CANADIAN COUNCIL OF RESOURCE MINISTERS
   1966. Pollution and our environment. Vols. 1, 2, and 3 and Conference Proceedings. Canadian Council of Resource Ministers, Montreal.

CONSERVATION COUNCIL OF ONTARIO
   1964. A report on water pollution in Ontario. Conservation Council of Ontario, Toronto.

COULTHARD, T. L., and STEIN, J. R.
   1969. A report on the Okanagan water investigation 1968-69. B.C. Dept. Lands, Forests and Water Resources, Water Resources Service, Victoria.

CROUTER, R. A., and VERNON, E. H.
   1959. Effects of black-headed budworm control on salmon and trout in British Columbia. Canadian Fish Culturist 24:23–40.

DAVIS, C. C.
   1964. Evidence for the eutrophication of Lake Erie from phytoplankton records. Limnology and Oceanography 9(3):275–83.

DAVY, F. B.; KLEEREKOPER, H.; and GENSLER, P.
   1972. Effects of exposure to sublethal DDT on the locomotor behavior of the goldfish (Carassius auratus). Journal of the Fisheries Research Board of Canada 29(9):1333–36.

deVOS, A.
   1966. Water pollution and recreational values. In Pollution and our environment, Vol. 1, Paper A4–1–4. Canadian Council of Resource Ministers, Montreal.

DIMOND, J. B.; GETCHELL, A. D.; and BLEASE, J. A.
   1971. Accumulation and persistence of DDT in a lotic ecosystem. Journal of the Fisheries Research Board of Canada 28(12):1877–82.

DYMOND, J. R., et al.
1952. Pollution of the Spanish River. A report to the special committee of the Research Council of Ontario. Ontario Department of Lands and Forests, Res. Rept. 25:1–106.

ELSON, P. F.
1967. Effects on wild young salmon of spraying DDT over New Brunswick forests. Journal of the Fisheries Research Board of Canada 24(4):731–67.

ELSON, P. E., and KERSWILL, C. J.
1967. Developing criteria for pesticide residues important to fisheries. Canadian Fisheries Reports 9:41–45.

FIMREITE, N.
1970. Mercury contamination of Canadian fish and fish-eating birds. Water and Pollution Control 108(11):21–26.

FISHERIES RESEARCH BOARD OF CANADA
1954. Effects of DDT spraying on fishes and aquatic insects. Annual Report, pp. 42–43.
1955. Effects of DDT spraying on Miramichi salmon and stream insects. Annual Report, pp. 47–49.
1956–57. Effects of DDT spraying. Annual Report, pp. 50–51.
1957–58. Manipulation of the salmon habitat. Annual Report, pp. 69–71.
1961–62. Effects of sublethal zinc and copper pollution. Annual Report, pp. 76–77.
1971. Journal of the Fisheries Research Board of Canada 28 (2). (J. C. Stevenson, ed.) (Entire volume devoted to "Experimental Lakes Area" of northwestern Ontario—a long-term program of the FRB Freshwater Institute at Winnipeg designed to study the effects of eutrophication.)

FREDEEN, F. J.; ARNASON, A. P.; BERCK, B.; and REMPEL, J. G.
1953. Further experiments with DDT in the control of Simulium arcticum Mall. in the North and South Saskatchewan Rivers. Canadian Journal of Agricultural Science 33:379–93.

FREDEEN, F. J. H.; SAHA, J. G.; and ROYER, L. M.
1971. Residues of DDT, DDE and DDD in fish in the Saskatchewan River after using DDT as a blackfly larvicide for twenty years. Journal of the Fisheries Research Board of Canada 28(1):105–9.

GEEN, G. H., and ANDREW, F. J.

1961. Limnological changes in Seton Lake resulting from hydroelectric diversions. International Pacific Salmon Fisheries Commission, Progress Report 8.

GOLDIE, C. A.

1967. Pollution and the Fraser. Report 1 Preliminary investigations of waste disposal to the lower Fraser River. B.C. Dept. Lands, Forests and Water Resources, Pollution Control Board, Victoria.

GORHAM, E., and GORDON, A. G.

1960. The influence of smelter fumes upon the chemical composition of lake waters near Sudbury, Ontario, and upon the surrounding vegetation. Canadian Journal of Botany 38(4):477–87.

GOURDEAU, J.-P.

1962. Industrial waste pollution of streams in Quebec. OWRC, Ontario Industrial Waste Conference, 9:203–19.

GRANT, C. D.

1967. Effects on aquatic insects of forest spraying with phosphamidon in New Brunswick. Journal of the Fisheries Research Board of Canada 24(4):823–32.

HARVEY, H. H., and COOPER, A. C.

1962. Origin and treatment of a supersaturated river water. International Pacific Salmon Fisheries Commission, Progress Report 9.

HATFIELD, C. T.

1969. Effects of DDT larviciding on aquatic fauna of Bobby's Brook, Labrador. Canadian Fish Culturist 40:61–72.

HEARNDEN, E. H.

1970. Mercury pollution: Fisheries Department acts quickly to safeguard public health. Fisheries of Canada 22(10):3–6.

IDE, F. P.

1954. Pollution in relation to stream life. OWRC, Ontario Industrial Waste Conference, 1:86–108.

1956. Effect of forest spraying with DDT on aquatic insects of salmon streams. Transactions of the American Fisheries Society 86:208–19.

1967. Effects of forest spraying with DDT on aquatic insects of salmon streams in New Brunswick. Journal of the Fisheries Research Board of Canada 24(4):769–805.

INTERNATIONAL JOINT COMMISSION (UNITED STATES AND CANADA)
  1965. Report on the pollution of Rainy River and Lake of the
        Woods. International Joint Commission, Ottawa.
  1968. Report on the pollution of the Red River. International
        Joint Commission, Ottawa.
  1970. Special report on potential oil pollution, eutrophication
        and pollution from watercraft. Third Interim Report on
        Pollution of Lake Erie, Lake Ontario and the
        International Section of the St. Lawrence River.
        International Joint Commission, Ottawa.
  1971. Pollution of Lake Erie, Lake Ontario, and the
        International Section of the St. Lawrence River.
        International Joint Commission, Ottawa.

INTERNATIONAL LAKE ERIE WATER POLLUTION BOARD and INTER-
    NATIONAL LAKE ONTARIO–ST. LAWRENCE RIVER WATER POL-
    LUTION BOARD (N. J. CAMPBELL, CHAIRMAN OF EDITORIAL COM-
    MITTEE)
  1969. Report to the International Joint Commission on the
        Pollution of Lake Erie, Lake Ontario, and the
        International Section of the St. Lawrence River. Vols. 1,
        2, and 3.

INTERNATIONAL PACIFIC SALMON FISHERIES COMMISSION
  1966. Effects of logging on the salmon and trout populations
        of the Stellako River. Progress Report 14. New
        Westminster, B.C.

JOHNSON, B. T.; SAUNDERS, C. R.; SANDERS, H. O.; and CAMPBELL, R. S.
  1971. Biological magnification and degradation of DDT and
        aldrin by freshwater invertebrates. Journal of the
        Fisheries Research Board of Canada 28(5):705–9.

JOHNSON, M. G.; MICHALSKI, F. P.; and CHRISTIE, A. E.
  1970. Effects of acid mine wastes on phytoplankton
        communities of two northern Ontario lakes. Journal of
        the Fisheries Research Board of Canada 27(3):425–44.

JONASSON, I. R.
  1970. Mercury in the natural environment: a review of recent
        work. Geological Survey of Canada, Paper 70–57.
        Department of Energy, Mines and Resources, Ottawa.

KEENLEYSIDE, M. H. A.

1959. Effects of spruce budworm control on salmon and other fishes in New Brunswick. Canadian Fish Culturist 24:17–22.

1967. Effects of forest spraying with DDT in New Brunswick on food of young Atlantic salmon. Journal of the Fisheries Research Board of Canada 24(4):807–22.

KELSO, J. R. M.; MacCRIMMON, H. R.; and ECOBICHON, D. J.

1970. Seasonal insecticide residue changes in tissues of fish from the Grand River, Ontario. Transactions of the American Fisheries Society 99(2):423–26.

KERSWILL, C. J.

1958. Effects of DDT spraying in New Brunswick on future runs of adult salmon. The Atlantic Advocate 48(8):65–68.

1967. Studies on effects of forest sprayings with insecticides 1952-63, on fish and aquatic invertebrates in New Brunswick streams: introduction and summary. Journal of the Fisheries Research Board of Canada 24(4):701–8.

KERSWILL, C. J., and EDWARDS, H. E.

1967. Fish losses after forest spraying with insecticides in New Brunswick, 1952-62, as shown by caged specimens and other observations. Journal of the Fisheries Research Board of Canada 24(4):709–29.

KUSSAT, R. H.

1969. A comparison of aquatic communities in the Bow River above and below sources of domestic and industrial wastes from the City of Calgary. Canadian Fish Culturist 40:3–31.

LARKIN, P. A.

1956. Power development and fish conservation on the Fraser River. Institute of Fisheries, University of B.C., Vancouver.

LARKIN, P. A.; ELLICKSON, P. J.; and LAURIENTE, D.

1970. The effects of toxaphene on the fauna of Paul Lake, British Columbia. Fisheries Management Publication No. 14, B.C. Fish and Wildlife Branch, Victoria.

LARKIN, P. A., and GRADUATE STUDENTS

1959. The effects on fresh water fisheries of man-made activities in British Columbia. Canadian Fish Culturist 25:27–59.

LEHMKUHL, D. M.
    1972. Change in thermal regime as a cause of reduction of
        benthic fauna downstream of a reservoir. Journal of the
        Fisheries Research Board of Canada 29(9):1329–32.

LI, M. F., and TRAXLER, G. S.
    1972. Tissue culture bioassay method for water pollution with
        special reference to mercuric chloride. Journal of the
        Fisheries Research Board of Canada 29:501–5.

LOCKHART, W. L.; UTHE, J. F.; KENNEY, A. R.; and MEHRLE, P. M.
    1972. Methylmercury in northern pike (Esox lucius):
        distribution, elimination, and some biochemical
        characteristics of contaminated fish. Journal of the
        Fisheries Research Board of Canada 29(11):1519–23.

LOFTHUS, K. H., and REGIER, H. A. (special editors)
    1972. Proceedings of the 1971 Symposium on Salmonid
        Communities in Oligotrophic Lakes. Journal of the
        Fisheries Research Board of Canada 29(6).

MacKAY, H. H.
    1930. Pollution problems in Ontario. Transactions of the
        American Fisheries Society 60:297–305.

MCKENZIE, R. A.
    1930. The reported decrease in fish life and the pollution of
        the Winnipeg River, Kenora, Ontario. Transactions of
        the American Fisheries Society 60:311–23.

MCKIM, J. M., and BENOIT, D. A.
    1971. Effects of long-term exposures to copper on survival,
        growth, and reproduction of brook trout (Salvelinus
        fontinalis). Journal of the Fisheries Research Board of
        Canada 28(5):655 62.

MCLAREN, R. E., and JACKSON, K. J.
    1966. The impact of water pollution on the uses for
        water-fisheries. In Pollution and our environment, Vol.
        1, Background Paper A4–1–2–1. Canadian Council of
        Resource Ministers, Montreal.

MacLENNAN, J. M.
    1954. Pollution and waterfowl. OWRC, Ontario Industrial
        Waste Conference, 1:109–14.

MAUCK, W. L., and COBLE, D. W.

   1971. Vulnerability of some fishes to northern pike (*Esox lucius*) predation. Journal of the Fisheries Research Board of Canada 28(7):957–69.

MENZIES, J. R.

   1961. Water pollution in Canada by drainage basins. *In* Resources for Tomorrow, Conference Background Papers, Vol. 1, pp. 353–64. Ottawa.

MILLER, R. B., and PAETA, M. J.

   1959. The effects of power, irrigation, and stock water developments on the fisheries of the South Saskatchewan River. Canadian Fish Culturist 25:13–26.

MUNRO, D. A., and SOLMAN, V. E. F.

   1966. The impact of water pollution on wildlife. *In* Pollution and our environment, Vol. 1, Background Paper A4–1–3. Canadian Council of Resource Ministers, Montreal.

NICHOLSON, L. J.

   1967. The Kootenays: environmental quality and waste management. *In* Proceedings of Conference on B.C. Environment, ed. H. M. Rosenthal, pp. 36–40.

ONTARIO DEPARTMENT OF ENERGY and RESOURCES MANAGEMENT

   1966. Raisin River Conservation Report, Toronto.

PETERSON, G. R.; WARREN, H. V.; DELAVAULT, R. E.; and FLETCHER, K.

   1970. Heavy metal content of some fresh water fishes in British Columbia. Fisheries Technical Circular 2. B.C. Fish and Wildlife Branch, Victoria.

PIPPY, J. H. C., and HARE, G. M.

   1969. Relationship of river pollution to bacterial infection in salmon *(Salmo salar)* and suckers *(Catostomus commersoni)*. Transactions of the American Fisheries Society 98(4):685–90.

PREVOST, G.

   1961. A typical drainage basin: the Ottawa River. *In* Resources for Tomorrow, Conference Background Papers, Supplementary Vol., pp. 83–111. Ottawa.

PRITCHARD, A. L.

  1959. The effects on fisheries of man-made changes in fresh water in the Maritime provinces. Canadian Fish Culturist 25:3–6.

RABBITS, F. T.; BANKS, G. N.; SIROIS, L. L.; and STEVENS, C. S.

  1971. Environmental control in the mining and metallurgical industries in Canada. National Advisory Committee on Mining and Metallurgical Research, Ottawa, January 1971.

RANDALL, A. P.

  1965. Evidence of DDT resistance in populations of spruce budworm, *Choristoneura fumiferana* (Clem.) from DDT-sprayed areas of New Brunswick. Canadian Entomologist 97(12):1281–93.

REINERT, R. E.; STEWART, D.; and SEAGRAN, H. J.

  1972. Effects of dressing and cooking on DDT concentrations in certain fish from Lake Michigan. Journal of the Fisheries Research Board of Canada 29(5):525–9.

RUDD, J. W. M., and HAMILTON, R. D.

  1972. Biodegradation of trisodium nitrilotriacetate in a model aerated sewage lagoon. Journal of the Fisheries Research Board of Canada 29(8):1203–8.

SAETHER, O. A.

  1970. A survey of the bottom fauna in lakes of the Okanagan Valley, British Columbia. Fisheries Research Board of Canada, Technical Report 196.

SAUNDERS, J. W.

  1969. Mass mortalities and behaviour of brook trout and juvenile Atlantic salmon in a stream polluted by agricultural pesticides. Journal of the Fisheries Research Board of Canada 26(3):695–99.

SAUNDERS, J. W., and SMITH, M. W.

  1965. Changes in a stream population of trout associated with increased silt. Journal of the Fisheries Research Board of Canada 22(2):395–404.

SAUNDERS, R. L., and SPRAGUE, J. B.

  1967. Effects of copper-zinc pollution on a spawning migration of Atlantic salmon. Water Research 1(5):419–32.

SCHERER, E.
1971. Effects of oxygen depletion and of carbon dioxide buildup on the photic behavior of the walleye *(Stizostedion vitreum vitreum)*. Journal of the Fisheries Research Board of Canada 28(9):1303–7.

SCHINDLER, D. W.; ARMSTRONG, F. A. J.; HOLMGREN, S. K.; and BRUN-SKILL, G. J.
1971. Eutrophication of lake 227, Experimental Lakes Area, northwestern Ontario, by addition of phosphate and nitrate. Journal of the Fisheries Research Board of Canada 28(11):1763–82.

SERVIZI, J. A., and GORDON, R. W.
1972. Detoxification of kraft pulp mill effluent by an aerated lagoon. International Pacific Salmon Fisheries Commission, Progress Report 26. New Westminster, B.C.

SERVIZI, J. A.; MARTENS, D. W.; and GORDON, R. W.
1970. Effects of decaying bark on incubation of salmon eggs. International Pacific Salmon Fisheries Commission, Progress Report 24. New Westminster, B.C.

SMALL, F. L.
1965. Industrial wastes and water pollution on the prairies. Water and Pollution Control 103(10):37–40.

SPRAGUE, J. B.
1964. Lethal concentrations of copper and zinc for young Atlantic salmon. Journal of the Fisheries Research Board of Canada 21(1):17–26.
1965. Effects of sublethal concentrations of zinc and copper on migration of Atlantic salmon. Biological Problems in Water Pollution. Third Seminar. U.S. Public Health Service Publication 999–WP–25, pp. 332–33.
1968. Apparent DDT tolerance in an aquatic insect disproved by test. Canadian Entomologist 100(3):279–84.

SPRAGUE, J. B., and CARSON, W. V.
1964. Chemical conditions in the northwest Miramichi River during 1963. Fisheries Research Board of Canada MS Report 180.

SPRAGUE, J. B., and DUFFY, J. R.
1971. DDT residues in Canadian Atlantic fishes and shellfishes in 1967. Journal of the Fisheries Research Board of Canada 28(1):59-64.

SPRAGUE, J. B., and RUGGLES, C. P.

    1966. Impact of water pollution on fisheries in the Atlantic provinces. *In* Pollution and our environment, Vol. 1, Background Paper A4–1–2–2. Canadian Council of Resource Ministers, Montreal.

SPRAGUE, J. B., and SAUNDERS, R. L.

    1963. Avoidance of sublethal mining pollution by Atlantic salmon. OWRC, Ontario Industrial Waste Conference, 10:221–36.

SPRAGUE, J. B.; ELSON, P. F.; and SAUNDERS, R. L.

    1965. Sublethal copper-zinc pollution in a salmon river—a field and laboratory study. International Journal of Air and Water Pollution 9(9):531–43.

TAYLOR, V. R.

    1965. Water pollution and fish populations in the province of Newfoundland and Labrador in 1964. Canadian Fish Culturist 35:3–15.

UTHE, J. F., and BLIGH, E. G.

    1971. Preliminary survey of heavy metal contamination of Canadian freshwater fish. Journal of the Fisheries Research Board of Canada 28(5):786–88.

VLADYKOV, VADIM D.

    1959. The effects on fisheries of man-made changes in fresh water in the province of Quebec. Canadian Fish Culturist 25:7–12.

WAVRE, M., and BRINKHURST, R. O.

    1971. Interactions between some tubificid oligochaetes and bacteria found in the sediments of Toronto Harbour, Ontario. Journal of the Fisheries Research Board of Canada 28(3):335–41.

WEBB, P. W., and BRETT, J. R.

    1972. The effects of sublethal concentrations of whole bleached kraftmill effluent on the growth and food conversion efficiency of underyearling sockeye salmon *(Oncorhynchus nerka)*. Journal of the Fisheries Research Board of Canada 29(11):1555–63.

WIMS, F. J.

  1962. Treatment of chrome-tanning wastes for acceptance by an activated sludge plant. OWRC, Ontario Industrial Waste Conference, 9:231–69.

WOBESER, G.; NIELSEN, N. O.; and DUNLOP, R. H.

  1970. Mercury concentrations in tissues of fish from the Saskatchewan River. Journal of the Fisheries Research Board of Canada 27(4):830–34.

ZITKO, V.

  1969. Chemical analysis of polluted water. CIC Pollution Conference, St. Mary's University, Halifax (E. R. Hayes, editor), pp. 132–39.

ZITKO, V.; FINLAYSON, B. J.; WILDISH, D. J.; ANDERSON, J. M.; and KOHLER, A. C.

  1971. Methylmercury in freshwater and marine fishes in New Brunswick, in the Bay of Fundy, and on the Nova Scotia banks. Journal of the Fisheries Research Board of Canada 28(9):1285–91.

ZYBLUT, E. R.

  1970. Long-term changes in the limnology and macrozooplankton of a large British Columbia lake. Journal of the Fisheries Research Board of Canada 27:1239–50.

INDEX

Acid mine wastes, 46–47
Activated sludge, 41
Aerial sprays, 56
Agricultural practices, 79, 87
Alaska, 29
Alberta, 27
Alkyl benzene sulphonate, 41
Alkyl mercury compounds, 51
Aluminum Company of
    Canada, 85
Amazon River, 27
Ammonia, 41, 64
Anadromous fish, 29
Anaerobic digestion, 41
Animal manure, 32, 43, 64
Annual crops of fish, 17
Anodonta, 13
Arctic char, 28
Arctic drainage, 29
Arctic grayling, 14
Arctic Ocean, 27
Arrow lakes, 26
Arsenic, 36, 64, 65
Artemia salina, 25
Athabasca River, 11, 19, 26,
    85
Atlantic salmon, 20, 28, 29, 64
Atton, Dr. M., 50
Atomic Energy of Canada
    Limited, 73

Babine Lake, B.C., 23

Bacteria: in lakes, 7; in
    methylation of mercury, 52;
    in sewage treatment, 40
Barren lakes, 22
Bass, 9, 22
Benzene, 65
Big Quill Lake, Sask., 25
Biochemical oxygen demand,
    40, 43
Black flies, 28, 58
Black liquor, 60
Bluegills, 10
Blue-green algae, 24, 35
Blue pike, 68
Bottom organisms, 8
Bras d'Or, Cape Breton
    Island, 23
British Columbia, 77
B.C. Pollution Control Board,
    69
Burbot (ling), 20

Caddisflies, 8
Cadmium, 33, 64
Canada Water Act, 89,
    Appendix (91)
Canadian Federation of
    Mayors and Municipalities,
    42
Canadian lakes and streams,
    15–30
Canadian Shield, 2, 15, 17–20,

27–28, 46; lakes of, 17–21
Cancer, 34, 73
Carbon, 34, 67
Carp, 63, 68
Catadromous fish, 29
Chironomid larvae, 39
Chlor-alkali plants, 50
Chlorination, 40–41
Christie Bay, Great Slave Lake, 19
Cirque lakes, 26
Ciscoes. *See* Lake herring
Cladophera, 39
Climate, influence of, on lakes, 3
Coal mines, 46
Columbia River, 83
Congo River, 27
Continental Divide, 27, 29
Copper, 33, 36
Copper sulphate, 58
Cornwall, Ont., 50
Cottids, 14
Crappies, 10
Cresols, 65
Cyanide, 48, 64

Dace, 14
Dams, 81–83, 85, 87; bypass for young salmon, 83; discharge orifices, 83; effect on downstream ecology, 84; effect on fish migration, 82–84; tailwater, 83
Damselflies, 8
DDE, 57
DDT, 33, 59; accumulation in fish eggs, 33, 57; effect on bird eggs, 58; effect on learning in salmon; 57, effect on moulting in insects, 57; effect on nesting behaviour in birds, 58; in Saskatchewan River, 58; persistence in soil, 57; storage in fatty tissue, 57
Detergents, 34, 42, 67, 87
Dissolved solids, 38

Dissolved oxygen, 5–6
Dragonflies, 8
Drum, 68

Eels, 29
Electro-plating wastes, 64
Enrichment of lakes and streams, 33, 38
Ephydra, 25
Epilimnion, 4–6, 72, 81
Estuaries, 84
Eutrophication, 34, 66–69, 77, 79
Eutrophic lakes, 10

Fertilizers, 34, 68, 79, 80
Fishflies, 8
Fishladders, 82–83
Fish locks, 82
Fish parasites, 30
Fishways, 82–83
Flin Flon, Man., 36
Fluorides, 65
Forestry practices, 87
Forestry wastes, 32, 43, 60
Fort MacMurray, N.W.T., 11
Fraser River, B.C., 27, 39, 82, 85
Fungi in sewage, 40
Fungicides, 79

Gammarus, 56
Garbage dumps, 69
Gaspe Peninsula, 16
Geological formation, influence of, on lakes, 2
Geosmin, 24
Glacial flour, 26
Glacial period, 29
Glacial relicts, 18
Goldeyes, 84
Goldfish, 63
Gravel removal, 79
Grease, 43, 53
Great Bear Lake, 15, 19, 86
Great Central Lake, 69
Great Circle Lakes (B.C.), 85
Great Lakes, 15, 27, 58, 77,

86, 88; as oligotrophic, 10;
chemical changes, 68;
fisheries, 68; history, 19;
mineral content, 20
Great Slave Lake, N.W.T.,
4–5, 15, 19–20, 84, 86
Greenhouses, 72
Gross pollutants, 45–47

Halifax, N.S., 40
Heavy metals, 33, 35, 47–49;
in fish tissues, 48
Herbicides, 58–59, 79
Hot water, 70–72
Hudson Bay, 27, 55
Hull, P.Q., 40
Human sewage, 67
Humic material, 10
Hydra, 13
Hydrogen sulphide, 52, 60
Hypolimnion, 4–5, 72

Incineration, 41
Inconnu, 28
Industrial effluents, 45
Industrial plants, 76
Insect larvae, 74
Insecticides, 79
Irrigation practices, 80, 87
Iron, 33

James Bay Hydroelectric
Power Development Project,
85–86

Kamloops trout, 56
Kemano River, B.C., 85
Kitimat, B.C., 85
Kootenay, Lake, B.C., 22, 26

Lake Athabasca, 15, 19–20,
84, 86
Lake Erie, 19, 50, 67–68, 87
Lake herring (ciscoes,
tullibee), 9, 18, 20, 22, 68
Lake Huron, 50, 68, 85
Lake Michigan, 20, 68
Lake Ontario, 20, 32, 67

Lake outlets, 13
Lake St. Clair, 50
Lake Superior, 2, 5, 20, 68
Lake Tanganyika, 5
Lake trout, 9, 17–18, 20, 22,
51
Lake Washington, 23
Lake Winnipeg, 15, 19–20, 50,
86
Lakes: aging of, 10; annual
cycle of, 4, 6, 8; as
ecosystems, 10; circulation
in, 72; depth as a factor in
productivity, 3, 5; dissolved
minerals in, 2, 21, 24, 26;
dissolved oxygen in, 5–6;
eutrophic, 10; ice-free
period, 4; influence of
topography, 2; of Alberta,
16, 24; of Atlantic
provinces, 16, 21; of British
Columbia, 16, 26; of
Labrador, 16; of Manitoba,
16, 24; of Ontario, 16; of
Quebec, 16; of
Saskatchewan, 16, 24;
oligotrophic, 10; physical
structure, 72; profundal
zone, 8; proglacial, 18;
seepage, 27; shape of basin,
2–3; stratification, 5
Lampreys, 14, 20
Land and water use, 76–90
Lead, 33
Lignin, 61
Limnological regions, 7, 27
Little Manitou Lake, 2
Logging, impact of, 76
Log "driving," 44, 78
Log jams, 78

Mackenzie drainage, 29
Mackenzie pipeline, 79
Mackenzie River, 11, 27–28
Manitoba, 16, 24, 36
Maritime provinces, 27, 47, 77
Mayflies, 13
Mercurial slimicides, 50

Mercury, 36, 49, 50; concentrations in fish, 52; discharge from cities, 51; forms of, 52; human consumption of, 51; uses of, 50–51
Mercury pollution, 49–53
Methyl mercaptans, 60
Migratory fishes, 82
Mine tailings, 48
Mining, 32, 47–48
Minnows, 9–10, 20, 22, 29
Mississippi drainage, 27, 29
Montreal, 39–40
Mountain lakes, 26–27
Mountain whitefish, 14
Mucous, precipitation on gills, 60
Muskeg, 17
Mussels, freshwater, 13
Mutation rates, 34, 73
Myoxocephalus quadricornus, 18
Mysis relicta, 18, 22

NAWAPA (North American Water and Power Alliance), 85
Nechako River, B.C., 85
Nelson River, Man., 27
Neomysis mercedis, 23
Nickel, 64
Nile River, 27
Nitrates, 33, 65, 69
Nitrogen bubble disease, 83
Northern Canada, 49
Northern suckers, 20
NTA (nitrylo-triacetate), 42
Nuclear power plants, 34, 68, 73–75
Nuclear wastes, 73–75

Oil, 53; as a human health hazard, 54; diesel fuel, 53; effects in Arctic, 54–55; effects on birds, 54; effects on fish, 53; effects on insects, 53; pipeline, 54–55; pollution, 53–55; sludge, 53; tankers, 55; waste treatment, 54
Okanagan Lake, B.C., 80
Ontario, 10, 16, 61
Organic matter, 38–39
Organic pollution, 32, 38–44
Organo-phosphorus insecticides, 56
Orinoco River, 27
Ottawa River, 50
Oxbow lakes, 11
Oxygen concentration, 59
Oxygen content of polluted streams, 32, 35, 38–39, 61
Oxygen demand, 60
Oxygen depletion, 67

Pacific salmon, 28–29; in the Great Lakes, 20, 57
Parkland lakes, 21–23
Paul Lake, B.C., 56
Peace River, B.C., 19–20, 84–85
Perch, 22, 68
Pesticides, 55, 80, 87; effects on fish, 56; in sewage, 41
pH: change from pollution, 63; effect on fish, 45–46
Phenol, 33, 63
Phenyl mercuric acetate, 50
Phosphates, 33, 35, 41–42, 67, 69, 87
Phosphorus, elemental, 65
Phytoplankton, 3, 7, 10, 52, 69, 74
Pickerel, 20
Pike, 9, 20, 22, 29, 30, 50, 51
Pinchi Lake, B.C., 36
Pollution: downstream effects, 32, 38–39; legislation, 89
Pontoporeia affinis, 18, 22–23
Powell Lake, B.C., 23
Power plants, 70–72
Precambrian Shield. *See* Canadian Shield
Primary production, 3
Prince Albert, Sask., 49

Prince Edward Island, 16
Prince Rupert, B.C., 40
Princeton, B.C., 69
Productivity, 74
Protozoa in sewage, 40
Pulp mills, 32, 49, 59; effluent, 33; kraft (sulphite) process, 60; soda process, 60
Pulp-mill wastes, 59–62; attraction of fish to, 61; effects on stream organisms, 61; effects on vegetation, 61; recycling, 62; pH of, 61; sub-lethal effects of, 61

Quaking bogs, 17
Quebec, 27
Quebec City, P.Q., 40
Quesnel Lake, B.C., 23

Radioactive wastes, 34, 73–74; concentration in organisms, 74
Rainbow trout, 22, 24, 30
Rawson, Dr. D. S., 1
Red blood cell counts, 61
Red Lake, N.W.T., 36
Red River, Man., 20
Redside shiners, 56
Reindeer Lake, 86
Rem, 73
Reservoirs, 81–82; as settling basins, 84; changes in water level, 81, 85; discharges from, 81; fish migration in, 83; flood control, 85; irrigation, 80; shoreline clearing, 85; physical circulation of, 81; productivity, 81
Resolute Bay, N.W.T., 69
Resource management, 88
Rhine River, 27, 64
Rivers. See Streams
Rivers and streams: Canadian, 27–30; general, 10–14
Road construction, 79
Rock slides, 82

Rocky Mountain Trench, 85
Rooted aquatic plants, 3, 8, 11; control of, 58; effects of pulp-mill wastes on, 61
Rotenone, 56
Run-off, 77

St. Lawrence River, 20, 27, 39, 42, 50, 87, 88
Saline lakes, 10, 23–25
Salinization (alkali), 80
Salmon, 14, 63, 76, 82–83, 85; chinook, 61; chum, 29; kokanee, 23; sockeye, 23, 69
Saskatchewan, 27
Saskatchewan River, 20, 50, 58
Sauger, 20, 68
Sawdust, 44
Seiches, 9
Septic tanks, 43, 69
Settling basins, 62
Sewage, 32, 35, 39, 41, 46, 72, 74, 76, 87; fungi, 38; grit chambers, 40; primary treatment, 40, 43; secondary treatment, 40; spray irrigation, 39–40; tertiary treatment, 41; treatment plants, 40, 72; trickling filter, 41
Shuswap Lake, B.C., 23
Silt, 77, 87
Similkameen River, B.C., 49
Slave River, N.W.T., 11, 19
Slime bacteria, 46
Sloughs, 24
Smelt, 68
Smelting wastes, 64
"Soap" lakes, 25
Sodium arsenite, 65
Sodium thiosulphate, 60
Soil erosion, 77, 79, 87
Spanish River, Ont., 61
Sphaerotilus, 38
Sphagnum moss, 17
Spillways, 83
Squawfish, 29

Steel mills, 34
Stoneflies, 13
Stream fishes, 13
Streams: as ecosystems, 14; as
    spawning sites for fish, 28;
    bottom organisms, 13;
    braided channels, 11; coastal
    areas, 12; "drift" organisms,
    13; fluctuations in level,
    11–12, 71; gradients, 11;
    mineral content of, 12, 63;
    mountain regions, 28;
    organic matter, 11–12;
    oxygen content, 12;
    parkland zone, 28,
    productivity, 11;
    temperatures, 12, 77; winter
    conditions, 28
Stuart Lake, 23
Sturgeon, 14
Sub-lethal effects, 48, 57, 64
Suckers, 9, 14, 22
Sudbury, Ont., 46
Sulphate bacteria, 25
Sulphur dioxide, 46
Suspended solids, 38
Swimmer's itch, 58

Tailings ponds, 48–49
"Thaw" lakes, 19
Thermal plants, 34
Thermocline, 4
TLM (Tolerance Limit
    Median), 34
Topography, influence of, on
    lakes, 2
Toronto, 32
Toxaphene, 56
Toxicants, 33
Tree farming, 78
Triaenophorus, 30
Trois Rivières, P.Q., 50
Trout, 14, 28, 63–64, 77
Trout Lake, B.C., 26
Trickling filter treatment, 41
2, 4, 5-T
    (trichlorophenoxyethanol),
    58–59

Tubificid worms, 39, 61

Vancouver, B.C., 40
Vegetable wastes, 32
Vernon, B.C., 40
Victoria, B.C., 40

W.A.C. Bennett Dam, 84
Walleye, 9, 22, 29, 51, 68
Waste heat (thermal
    pollution), 34, 35, 70–74; in
    lakes, 34, 71
Water diversions, 85–86
Waterton Lake, Alta., 22
Watrous, Sask., 2
Welland Canal, 20
Whitefish, 9, 17–18, 20, 22,
    25, 51, 68
Wind, effect of, on lake
    structure, 4, 9
Winter kill, 6, 24

Yorkton, Sask., 25
Yukon, 29
Yangtse River, 27

Zinc, 36, 64
Zooplankton, 7, 10, 69